# RESERVOIR ENGINEERING
# TECHNIQUES USING FORTRAN

# RESERVOIR ENGINEERING TECHNIQUES USING FORTRAN

**MIHIR K. SINHA, Ph.D.**
Rocky Mountain Petroleum Consultants
Salt Lake City, Utah

**LARRY R. PADGETT, Ph.D.**
W.V.N.E.T.
Morgantown, W. Va.

D. Reidel Publishing Company
A Member of the Kluwer Academic Publishers Group
Dordrecht/Boston/Lancaster

International Human Resources Development Corporation • Boston

**Library of Congress Cataloging in Publication Data**

Sinha, Mihir K., 1941–
  Reservoir engineering techniques using Fortran.

  Bibliography: p.
  Includes index.
  1. Oil reservoir engineering—Computer programs.
  2. FORTRAN (Computer program language) I. Padgett,
Larry R., 1936–  . II. Title.
TN871.S554   1984    622'.3382     84–16106
ISBN-13: 978-94-010-8837-4     e-ISBN-13: 978-94-009-5293-5
DOI: 10.1007/978-94-009-5293-5

Published by D. Reidel Publishing Company
P.O. Box 17, 3300 AA Dordrecht, Holland in co-publication with IHRDC

Sold and distributed in North America by IHRDC

In all other countries, sold and distributed by Kluwer Academic Publishers Group, P.O. Box 322, 3300 AH Dordrecht, Holland

To Abanish Chandra Sinha, Encouraging and Supportive Father,
and
Linda Padgett, Understanding Wife

# CONTENTS

Preface    ix

Introduction    xi


## I RESERVOIR FLUID PROPERTIES, HYDROCARBONS IN PLACE, AND RESERVES

1 Physical Properties of Reservoir Hydrocarbon Fluids    3

2 Relationship Between Permeability Ratio ($k_g/k_o$) and Total Liquid Saturation ($S_l$) for a Reservoir    13

3 Conventional Well Log Analysis    19

4 Oil in Place and Recoverable Reserve by the Volumetric Method    31

5 Estimation of Initial Oil in Place by the Material Balance Method for a Solution Gas Drive Reservoir    41

6 Determination of Original Oil in Place by the Material Balance Method for a Reservoir with Initial Gas Cap and No Water Influx    47

7 Determination of Oil in Place by the Material Balance Method for Reservoirs with Partial Water Drive (No Gas Cap)    55


## II FUTURE RESERVOIR PERFORMANCE (OIL)

8 Performance Prediction by Production Decline Analysis    71

9 Prediction of Performance and Ultimate Oil Recovery of a Combination Solution Gas/Gas-Cap Drive Reservoir    81

10 Prediction of Performance of a Reservoir with Partial Edge-Water Drive    97

11 Dispersed Gas Injection Performance    113

III   ENHANCED OIL RECOVERY AND PERFORMANCE
      BY EMPIRICAL METHODS

12   In Situ Combustion Performance Using the Oil-Displaced/Volume-Burned
     Method                                                                    127

13   In Situ Combustion Performance Using Empirical Correlations               135

14   Carbon Dioxide Flood Performance                                          141

15   Polymer Flood Performance                                                 147

IV   RESERVOIR ENGINEERING FOR NATURAL GAS

16   Physical Properties of Natural Gas                                        155

17   Determination of Gas in Place by the Material Balance Method for a Water
     Drive Reservoir                                                           163

18   Determination of Original Gas in Place for an Abnormally Pressured Reservoir  175

19   Static/Flowing Bottomhole Pressure for a Gas Well                         183

20   Stabilized Absolute Open Flow Potential of a Gas Well                     191

21   Conversion of Point-after-Point Gas Well Test Results to Equivalent
     Isochronal Test Results                                                   201

22   Gas Well Deliverability                                                   209

     Index                                                                     221

# PREFACE

Practical reservoir engineering techniques have been adequately described in various publications and textbooks, and virtually all useful techniques are suitable for implementation on a digital computer. Computer programs have been written for many of these techniques, but the source programs are usually not available in published form.

The purpose of this book is to provide a central source of FORTRAN-coded algorithms for a wide range of conventional reservoir engineering techniques. The book may be used as a supplementary text for courses in practical reservoir engineering. However, the book is primarily intended for practicing reservoir engineers in the hope that the collection of programs provided will greatly facilitate their work.

In addition, the book should be also helpful for non–petroleum engineers who are involved in applying the results of reservoir engineering analysis.

Sufficient information is provided about each of the techniques to allow the book to be used as a handy reference.

# INTRODUCTION

This book provides many of the useful practical reservoir engineering (conventional) techniques used today in the form of FORTRAN codes. The primary objectives have been to provide the simplest possible method for obtaining reliable answers to practical problems. Unfortunately, these codes can usually be applied by simply following a cookbook approach. However, if at all possible, the solutions obtained should be verified and cross-checked by some other means and, most important, should be checked for reasonability.

Working equations have been provided to help or facilitate the understanding of the basic approach. Example problems have been solved using the codes. Although applications are emphasized in the book, enough theoretical concepts are included to give the fundamentals of the methods used. Bibliographies are included to give additional material for study. No effort has been made to cover everything that is known on a particular subject or to reference all useful information.

There are 22 FORTRAN codes covering a wide range of conventional reservoir engineering techniques. These range from physical properties of petroleum fluids to gas well deliverability to enhanced oil recovery. However, numerical reservoir simulation models are not included. For more detailed and rigorous reservoir studies, it may be necessary to use sophisticated numerical reservoir simulation models to gain better insight into finer aspects of reservoir mechanics.

# I RESERVOIR FLUID PROPERTIES, HYDROCARBONS IN PLACE, AND RESERVES

# 1  PHYSICAL PROPERTIES OF RESERVOIR HYDROCARBON FLUIDS

## PURPOSE

The program is designed to provide estimates of physical properties of reservoir fluids (hydrocarbons) as a function of pressure based on the published empirical correlations. Most of the desired properties can be correlated as functions of pressure, temperature, oil gravity, and gas gravity. The program provides quick and reliable answers with no need for a time-consuming and tedious process of reading graphs, monograms, or charts and tables.

The evaluation of physical properties of reservoir fluids (hydrocarbons) is an important first step for any reservoir engineering study. Laboratory measurement of the physical properties, if available, should be used for such studies. However, most often such measured values are not available. Furthermore, it is becoming increasingly necessary to have accurate estimates of physical properties of reservoir fluids in advance of laboratory pressure-volume-temperature (PVT) studies to allow management to review project feasibility studies. Therefore, a simple program like this one will be helpful to practicing engineers in obtaining such information for use in various analytical and numerical methods of reservoir evaluation.

## METHOD

$$\gamma_{gs} = \gamma_{gp} \left[1 + 5.912 \times 10^{-5}\gamma_o T \log(p/114.7)\right]$$
$$R_s = C_1\gamma_{gs}p^{C2} \exp\{C_3[\gamma_o/(T + 460)]\}$$
$$B_o = 1 + C_4 R_s + C_5(T - 60)(\gamma_o/\gamma_{gs}) + C_6 R_s(T - 60)(\gamma_o/\gamma_{gs}) \text{ when } p \le p_b$$
$$B_o = B_{ob} \exp[C_o(p - p_b)] \text{ when } p > p_b$$
$$C_o = (C_7 + C_8 R_s + C_9 + T + C_{10}\gamma_{gs} + C_{11}\gamma_o)/C_{12}p$$
$$\mu_o = \mu_{ob}(p/p_b)^m$$
$$m = C_{13}p^{C14} \exp(C_{15} + C_{16}p)$$

$$ZZ = 3.0324 - 0.02023\gamma_o$$
$$Y = 10^{ZZ}$$
$$X = YT^{-1.163}$$
$$\mu_{OD} = 10^X - 1$$
$$A = 10.715(R_s + 100)^{-0.515}$$
$$B = 5.44(R_s + 150)^{-0.338}$$
$$\mu = A\mu_{OD}^B$$
$$Z = (1 + y + y^2 - y^3)/(1 - y^3)^{-(14.76t - 9.76t^2 + 4.58t^3)y} + (90.7t - 242.2t^2 + 42.4t^3)\, y^{\,(1.18 + 2.82t)}$$
$$y = bp/4$$
$$bp_c/RT_c = 0.245 \exp[-1.2(1 - t)^2]$$

where

| Constant | Values | |
|---|---|---|
| | $\gamma_o \le 30$ | $\gamma_o > 30$ |
| $C_1$ | 0.0362 | 0.0178 |
| $C_2$ | 1.0937 | 1.1870 |
| $C_3$ | 25.724 | 23.9310 |
| $C_4$ | $4.677 \times 10^{-4}$ | $4.670 \times 10^{-4}$ |
| $C_5$ | $1.751 \times 10^{-5}$ | $1.100 \times 10^{-5}$ |
| $C_6$ | $-1.811 \times 10^{-8}$ | $1.337 \times 10^{-9}$ |
| $C_7$ | $-1433.0$ | $-1433.0$ |
| $C_8$ | 5.0 | 5.0 |
| $C_9$ | 17.2 | 17.2 |
| $C_{10}$ | $-1180.0$ | $-1180.0$ |
| $C_{11}$ | 12.61 | 12.61 |
| $C_{12}$ | $10^5$ | $10^5$ |
| $C_{13}$ | 2.6 | 2.6 |
| $C_{14}$ | 1.187 | 1.187 |
| $C_{15}$ | $-11.513$ | $-11.513$ |
| $C_{16}$ | $-8.98 \times 10^{-5}$ | $-8.98 \times 10^{-5}$ |

$B_o$ = Oil formation volume factor, bbl/STB;

$B_{ob}$ = Oil formation volume factor at $p_b$, bbl/STB;

$C_o$ = Oil compressibility, vol/vol/psi;

$p$ = Pressure, psia;

$p_b$ = Bubble point pressure, psia;

$R_s$ = Solution gas-oil ratio, SCF/STB;

$\gamma_g$ = Gas gravity (air = 1);

$\gamma_{gs}$ = Gas gravity at separator pressure of 100 psig;

$\gamma_{gp}$ = Gas gravity at separator pressure $P$ and temperature $T$;

$P$ = Actual separator pressure, psia;

$T$ = Actual separator temperature, °F;

$\gamma_o$ = Oil gravity, °API;

$\mu_o$ = Oil viscosity, cp;

$\mu_{ob}$ = Oil viscosity at bubble point, cp;

$\mu_{oD}$ = Dead oil viscosity at $T$, cp;

$\mu$ = Viscosity of gas-saturated oil at $T$, cp;

$t$ = Reciprocal reduced temperature, °R.

"Gas gravity is a very strong correlating parameter and, unfortunately is one of the parameters of most questionable accuracy, because it depends on conditions at which the gas/oil separation is made" (Vazquez and Beggs 1980, 969). Therefore, caution should be exercised while utilizing this program to input very accurately the values of gas gravity and gravity measurement pressure and temperature.

**PROGRAM DESCRIPTION**

The FORTRAN program consists of a main program and two subprograms. The main program computes most of the oil physical properties, while the two subprograms, COMFAC and VISCO, provide gas compressibility factor ($Z$) and gas viscosity ($\mu_g$) values respectively.

The program will generate values for (1) the oil formation volume factor ($B_O$), (2) the solution gas-oil ratio (RS), (3) the gas formation volume factor (BG), (4) the oil viscosity (VISO), (5) the gas viscosity (VISG), (6) the two-phase formation volume factor (BT), and (7) the gas compressibility factor (Z-FAC) for each pressure point desired. Up to 50 pressure points can be input. Gas compressibilities are computed in the pseudo-reduced temperature range between 1.2 and 3 and the pseudo-reduced pressure value of less than 20.

**Input Format**

| Card | Format | Content |
|---|---|---|
| 1 | 2OA4 | AA |
| 2 | I2 | N |
| 3 | F5.3, F5.1, 6F5.0 | YGP, YOIL, SP, ST, PRI, PB, TRE, |
|  | F5.1, 3F5.4 | CP, CT, XCO2, XN2, XH2S |
| 4–N | F1$\phi$.1 | CI |

**Description of Input Parameters**

AA(1) = Uses up to 80 alphanumeric characters for project identification and description—e.g.,
Example Problem 1    Morgantown    West Virginia;

     N = Number of pressure points at which physical properties are to be eval-
         uated—e.g., 12;
   YGP = Gas gravity at separator pressure and temperature (air = 1);
  YOIL = Oil gravity, °API;
    SP = Separator pressure, psia;
    ST = Separator temperature, °F;
   PRI = Initial reservoir pressure, psi;
    PB = Bubble point pressure, psi;
   TRE = Reservoir temperature, °F;
    CP = Critical pressure (gas), psia;
    CT = Critical temperature (gas), °R;
  $XCO_2$ = Mol fraction of carbon dioxide in the gas;
   $XN_2$ = Mol fraction of nitrogen in the gas;
  $XH_2S$ = Mol fraction of hydrogen sulfide in the gas;
  C1(I) = Pressures at which the properties are to be computed; should corre-
         spond with N—that is, 12 pressure values must be input in cards 4
         through 15.

**EXAMPLE PROBLEM**

Compute the reservoir fluid properties at the following reservoir pressures 6000,
5000, 4000, 1900, 1700, 1500, 1300, 1100, 900, 700, 500, and 300 psia. Other
properties are as follows:

Gas gravity at separator pressure and temperature = 0.7 (air = 1),
Oil gravity = 35°API,
Separator pressure = 100 psia,
Separator temperature = 60°F,
Initial reservoir pressure = 6000 psia,
Bubble point pressure = 4000 psia,
Reservoir temperature = 220°F,
Critical pressure (gas) = 670 psia,
Critical temperature (gas) = 410°R,
Mol fraction of carbon dioxide in the gas = 0,
Mol fraction of nitrogen in the gas = 0,
Mol fraction of hydrogen sulfide in the gas = 0.

**INPUT LISTING**

FORTRAN CODING FORM

UU/CC   NAME _____   DATE _____

PROGRAM _____
ROUTINE _____

0 = ZERO       1 = ONE        2 = TWO
Ø = ALPHA O    I = ALPHA I    Z = ALPHA Z

| CARD | STATEMENT NUMBER (1-5) | 6 | 7,8,9,... (columns 7-60) |
|------|------------------------|---|--------------------------|
| 1 | EXAMPL | E | PROBLEM 1        MORGANTOWN        WEST VIRGINIA |
| 2 | 12 | | |
| 3 | -7.00 | | 3.5.0  4.80  6.0  60.00  40.00  2.20  6.70  4.10.0.0.0  .0  .0 |
| 4-1 | 6 | | 0.0 |
| 4-2 | 5 | | 0.0 |
| 4-3 | 4 | | 0.0 |
| 4-4 | 12 | | 0.0 |
| 4-5 | 17 | | 0.0 |
| 4-6 | 15 | | 0.0 |
| 4-7 | 13 | | 0.0 |
| 4-8 | 11 | | 0.0 |
| 4-9 | 9 | | 0.0 |
| 4-10 | 7 | | 0.0 |
| 4-11 | 5 | | 0.0 |
| 4-12 | 3 | | 0.0 |

## SOURCE LISTING

```
      THIS PROGRAM IS DESIGNED TO COMPUTE PVT PROPERTIES OF RESERVOIR
      HYDROCARBON FLUID UTILIZING STANDARD CORRELATIONS.  THE FOLLOWING
      INFORMATION MUST BE PROVIDED AS INPUT DATA.
    1 GAS GRAVITY
    2 PRESSURE AT WHICH GAS SAMPLE WAS TAKEN, SUCH AS SEPARATOR PRESSURE
    3 TEMPERATURE AT WHICH GAS SAMPLE WAS TAKEN, FAHRENHEIT
    4 OIL GRAVITY, API
    5 INITIAL RESERVOIR PRESSURE, PSI
    6 BUBBLE-POINT PRESSURE, PSI
    7 RESERVOIR TEMPERATURE, FAHRENHEIT


      DIMENSION C1(50),C2(50),C3(50),C4(50),C5(50),C6(50),C7(50),
     *AA(20),C8(50)
      C01=0.0362
      C02=1.0937
      C03=25.724
      C51=0.0178
      C52=1.187
      C53=23.931
      B1=4.677E-04
      B2=1.751E-05
      B3=-1.811E-08
      B11=4.67E-04
      B12=1.1E-05
      B13=1.337E-09
      A11=-1433.
      A21=5.
      A31=17.2
      A41=-1180.
      A51=12.6
      A61=1.E05
      D1=2.6
      D2=1.187
      D3=-11.513
      D4=-8.98E-05

      N=NUMBER OF PRESSURE VALUES TO BE READ

      READ(5,5) (AA(I),I=1,20)
    5 FORMAT(20A4)
      READ(5,1)N
    1 FORMAT(I2)

      READ BASIC OIL AND GAS DATA TO BE USED FOR CALCULATING PVT DATA

      READ(5,2)YGP,YOIL,SP,ST,PRI,PB,TRE,CP,CT,XCC2,XN2,H2S
    2 FORMAT(F5.3,F5.1,6F5.0,F5.1,3F5.4)
      YGS=YGP*(1.+5.912E-05*YOIL*ALOG10(SP/114.7))

      READ PRESSURES AT WHICH PVT PROPERTIRDES ARE DESIRED, N TIMES

      DO 107 I=1,N
      READ(5,3)C1(I)
    3 FORMAT(F10.1)
      IF(YOIL.GT.30.) GO TO 101
      C3(I)=C01*YGS*(C1(I)**C02)*EXP(C03*(YOIL/(ST+460.)))
      GO TO 102
  101 C3(I)=C51*YGS*(C1(I)**C52)*EXP(C53*(YOIL/(ST+460.)))
  102 IF(C1(I).GT.PB) GO TO 104
      IF(YOIL.GT.30.) GO TO 103
      C2(I)=1.+B1*C3(I)+B2*(ST-60.)*(YOIL/YGS)+B3*C3(I)*(ST-60.)
     **(YOIL/YGS)
      GO TO 104
  103 C2(I)=1.+B11*C3(I)+B12*(ST-60.)*(YOIL/YGS)+B13*(ST-60.)
     **(YOIL/YGS)
  104 CONTINUE
      IF(YOIL.GT.30.) GO TO 105
      RSB=C01*YGS*(PB**C02)*EXP(C03*(YOIL/(ST+460.)))
      GO TO 106
  105 RSB=C51*YGS*(PB**C52)*EXP(C53*(YOIL/(ST+460.)))
  106 IF(C1(I).GT.PB)BOB=1.+B11*RSB+B12*(ST-60.)*(YOIL/YGS)
     *+B13*(ST-60.)*(YOIL/YGS)
      CO=(A11+A21*C3(I)+A31*ST+A41*YGS+A51*YOIL)/(A61*C1(I))
      IF(C1(I).GE.PB)C2(I)=BOB*EXP(-CO*(C1(I)-PB))
      IF(C1(I).GE.PB)C3(I)=RSB
  107 CONTINUE

      CALCULATE OIL VISCOSITY
```

```
      Z=3.0324-.02023*YCIL
      Y=10.**Z
      X=Y*TRE**(-1.163)
      MOD=10.**X-1.
      DO 6000 I=1,N
      A=10.715*(C3(I)+100.)**(-0.515)
      B=5.44*(C3(I)+150.)**(-0.338)
      IF(C1(I).LE.PB)C5(I)=A*MOD**B
6000  CONTINUE
      A=10.715*(RSB+100.)**(-.515)
      B=5.44*(RSB+150.)**(-.338)
      VOB=A*MOD**B
      SM=D1*(C1(I)**D2)*EXP(D3+D4*C1(I))
      DO 108 I=1,N
      IF(C1(I).GT.PB)C5(I)=VOB*(C1(I)/PB)**SM
108   CONTINUE
      DO 109 I=1,N
      PR=C1(I)
      CALL COMFAC(PR,TRE,CP,CT,Z)
      CALL VISCO(PR,TRE,CP,CT,YGP,XCO2,XN2,H2S,V)
      TRA=TRE+460.
      C4(I)=0.00504*Z*TRA/C1(I)
      C7(I)=C2(I)+C4(I)*(C3(1)-C3(I))
      C6(I)=V
      C8(I)=Z
109   CONTINUE
      WRITE(6,6)(AA(I),I=1,20)
6     FORMAT(///,3X,20A4)
      WRITE(6,30)
30    FORMAT(//,5X,'PRESS',4X,'BO',4X,'RS',8X,'BG',7X,'VISO',5X,'VISG',
     *6X,'BT',6X,'Z-FAC')
      DO 4000 I=1,N
      WRITE(6,32)C1(I),C2(I),C3(I),C4(I),C5(I),C6(I),C7(I),C8(I)
32    FORMAT(/,2X,F8.0,3X,F4.2,1X,F6.0,3X,F8.5,3X,F6.3,3X,F6.5,3X,F5.1,
     *5X,F6.4)
4000  CONTINUE
      END
      SUBROUTINE COMFAC(T,S,PC,TC,Z)
      PR=T
      X=S+460.
      RRT=TC/X
      RP=PR/PC
      A=0.06125*RRT*EXP(-1.2*(1.-RRT)**2)
      B=RRT*(14.76-9.76*RRT+4.58*RRT**2)
      C=RRT*(90.7-242.2*RRT+42.4*RRT**2)
      D=2.18+2.82*RRT
      Y=0.01
      DO 2 J=1,30
      IF(Y.GT.1.)Y=0.6
      F=-A*RP+Y*(1.+(Y*(1.+Y*(1.-Y))))/(1.-Y)**3-B*Y**2+C*Y**D
      IF(ABS(F)-1.0E-6)4,4,3
3     DFDY=(1.+4.*Y*(1.+Y*(1.-Y))+Y**4)/(1.-Y)**4-2.*B*Y+D*C*Y**(D-1.)
2     Y=Y-F/DFDY
4     Z=A*RP/Y
      RETURN
      END
      SUBROUTINE VISCO(PB,TRE,PC,TC,GG,C1,C2,C3,VIS)
      REAL M
      DATA X0,X1,X2,X3,X4,X5,X6,X7,X8/1.112391E-2,1.677266E-5,2.113605E
     *-9,-1.094850E-4,-6.403164E-8,8.993745E-11,4.577352E-7,2.129034E-1
     *0,3.977322E-13/
      DATA A0,A1,A2,A3,A4,A5,A6,A7,A8,A9/-2.462118,2.970527,-2.862641E-
     *1,8.054205E-3,2.808609E-3,-3.498033,3.603730E-1,-1.044324E-2,-7.93385
     *7E-1,1.396433/
      DATA Z0,Z1,Z2,Z3,Z4,Z5/-1.491449E-1,4.410155E-3,8.393872E-2,-1.86
     *4088E-1,2.033679E-2,-6.095793E-4/
      PR=PB/PC
      TRA=TRE+460.
      TR=TRA/TC
      IF(TR.GT.3.)GO TO 20
      IF(TR.LT.1.2)GO TO 30
      IF(PR.GT.20.) GO TO 40
      M=28.97*GG
      Y5=GG/(39.8782+64.0537*GG)
      Y6=1.03401E-2-3.4823E-3/GG
      Y7=8.08098E-3-3.99988E-3/GG
      U=TRE
      V1=C1*Y5+C2*Y6+C3*Y7
      V2=X0+X1*U+X2*U**2+X3*M+X4*U*M+X5*U**2*M+X6*M**2+X7*U*M**2+X8*U**
     *2*M**2
      V3=((V1+V2)*1.E05+0.5)*1.E-5
      IF(PR.LT.1.)GO TO 50
      V4=A0+A1*PR+A2*PR**2+A3*PR**3+TR*(A4+A5*PR+A6*PR**2+A7*PR**3)
      V5=TR**2*(A8+A9*PR+Z0*PR**2+Z1*PR**3)
      V6=TR**3*(Z2+Z3*PR+Z4*PR**2+Z5*PR**3)
      V7=V4+V5+V6
      V8=EXP(V7)/TR
      U1=((V3*V8)*1.E04+0.5)*1.E-4
```

```
        GO TO 65
   20 WRITE(6,21)
   21 FORMAT(//,3X,'PSEUDO REDUCED TEMPERATURE IS GREATER THAN 3.0')
   30 WRITE(6,31)
   31 FORMAT(//,3X,'PSEUDO REDUCED TEMPERATURE IS LESS THAN 1.2')
   40 WRITE(6,41)
   41 FORMAT(//,3X,'PSEUDO REDUCED PRESSURE IS GREATER THAN 20')
        GO TO 60
   50 U1=V3
        GO TO 65
   60 U1=0.
   65 CONTINUE
      VIS=U1
      RETURN
      END
//GO.SYSIN DD *
 EXAMPLE PROBLEM 1  MORGANTOWN  WEST VIRGINIA
12
 .700 35.0 100.  60.6000.4000. 220. 670.410.
      6000.
      5000.
      4000.
      1900.
      1700.
      1500.
      1300.
      1100.
       900.
       700.
       500.
       300.
/*
```

## OUTPUT LISTING

EXAMPLE PROBLEM 1   MORGANTOWN   WEST VIRGINIA

| PRESS | BO | RS | BG | VISO | VISG | BT | Z-FAC |
|-------|------|-------|---------|-------|--------|------|--------|
| 6000. | 1.51 | 1177. | 0.00061 | 0.272 | .03120 | 1.5 | 1.0676 |
| 5000. | 1.53 | 1177. | 0.00067 | 0.271 | .02781 | 1.5 | 0.9805 |
| 4000. | 1.55 | 1177. | 0.00078 | 0.269 | .02432 | 1.5 | 0.9048 |
| 1900. | 1.23 | 486.  | 0.00154 | 0.402 | .01716 | 2.3 | 0.8531 |
| 1700. | 1.20 | 426.  | 0.00174 | 0.425 | .01652 | 2.5 | 0.8608 |
| 1500. | 1.17 | 367.  | 0.00199 | 0.452 | .01588 | 2.8 | 0.8708 |
| 1300. | 1.14 | 310.  | 0.00233 | 0.484 | .01525 | 3.2 | 0.8829 |
| 1100. | 1.12 | 254.  | 0.00279 | 0.521 | .01464 | 3.7 | 0.8970 |
| 900.  | 1.09 | 200.  | 0.00348 | 0.568 | .01403 | 4.5 | 0.9129 |
| 700.  | 1.07 | 149.  | 0.00455 | 0.626 | .01343 | 5.8 | 0.9302 |
| 500.  | 1.05 | 100.  | 0.00650 | 0.700 | .01272 | 8.1 | 0.9489 |
| 300.  | 1.03 | 54.   | 0.01107 | 0.800 | .01272 | 13.4 | 0.9687 |

## BIBLIOGRAPHY

Beggs, H. Dale, and Robinson, J. F.: "Estimating the Viscosity of Crude Oil Systems," *J. Pet. Tech.* (September 1975) 1140–1141.

Hall, Kenneth, and Yarborough, Lyman: "A New Equation of State for Z-Factor Calculations," *Oil and Gas J.* (June 18, 1973) 82–92.

Vazquez, Milton, and Beggs, H. Dale: "Correlations for Fluid Physical Property Prediction," *J. Pet. Tech.* (June 1980).

Yarborough, Lyman, and Hall, K. R.: "How to Solve Equation of State for Z-Factor," *Oil and Gas J.* (February 18, 1974) 86–88.

# 2 RELATIONSHIP BETWEEN RELATIVE PERMEABILITY RATIO ($k_g/k_o$) AND TOTAL LIQUID SATURATION ($S_l$) FOR A RESERVOIR

**PURPOSE**

This program is designed to determine the gas-oil relative permeability ratio ($k_g/k_o$) and the corresponding liquid saturation ($S_l$) relationships from actual production performance and fluid property data. The method is applicable for a volumetric, solution gas drive reservoir. The theoretical development of the procedure assumes that the reservoir pore volume remains constant during the producing history; that is, no water or gas influx takes place. All computations are made below the bubble point pressure.

In spite of various other restrictive assumptions that are made in the development of the theoretical basis and the limitation imposed by actual field production practices, it is believed that the permeability ratio–liquid saturation relationship developed by this method is a good and acceptable representation of the reservoir behavior.

**METHOD**

Instantaneous gas-oil ratio is used to determine the relative permeability ratio and is given by

$$\frac{k_g}{k_o} = (R_I - R_s) \times \frac{\mu_g}{\mu_o} \times \frac{B_g}{B_o}. \tag{2.1}$$

Corresponding total liquid saturation is given by

$$S_1 = S_w + (1 - S_w) \times \frac{N - N_P}{N} \times \frac{B_o}{B_{oi}}, \tag{2.2}$$

where

$R_I$ = Instantaneous producing gas-oil ratio, SCF/STB;

13

$R_S$ = Solution gas-oil ratio, SCF/STB;

$\mu_o$ = Oil viscosity at reservoir condition, cp;

$\mu_g$ = Gas viscosity at reservoir condition, cp;

$B_o$ = Oil formation volume factor, bbl/STB;

$B_g$ = Gas formation volume factor, bbl/SCF;

$k_g/k_o$ = Gas-oil relative permeability ratio;

$S_l$ = Total liquid saturation, fraction of pore volume;

$S_w$ = Interstitial water saturation, fraction of pore volume;

$N$ = Initial oil-in-place, STB;

$N_p$ = Cumulative oil production, STB;

$B_{oi}$ = Initial oil formation volume factor, bbl/STB.

## PROGRAM DESCRIPTION

The program consists of a simple FORTRAN main program, which will handle a maximum of 25 production and fluid property data. The procedure involves step-by-step computation of equations (2.1) and (2.2) for each pressure production step.

### Input Formats

| Card | Format | Content |
|------|--------|---------|
| 1 | 20A4 | A(I) |
| 2 | I2, 2F5.3, F10.0, F6.0 | N, SWI, BOB, OOIP, BPP |
| 3–N | 2F6.0, F5.3, F8.6, F6.0, F10.0, F5.2, F10.0 | Cl(I), C2(I), C3(I), C4(I), C5(I), C6(I), C7(I), C18(I) |

### Description of Input Parameters

A(I)    = Up to 80 alphanumeric characters are used for project identification;

N       = Number of input production–fluid property data set provided;

SWI   = Interstitial water saturation, fraction of pore space;

BOB  = Bubble point oil formation volume factor, bbl/STB;

OOIP = Original oil in place, STB;

BPP   = Bubble point pressure, psi;

C1(I)  = Average reservoir pressure, psi;

C2(I)  = Instantaneous producing gas-oil ratio, SCF/STB;

C3(I)  = Oil formation volume factor, bbl/STB;

C4(I)  = Gas formation volume factor, bbl/SCF;

C5(I)  = Solution gas-oil ratio, SCF/STB;

C6(I)  = Cumulative oil production, STB;
C7(I)  = Oil viscosity, cp;
$C18(I)$ = Gas viscosity, cp.

**EXAMPLE PROBLEM**

The production and fluid property data for a sand reservoir are given in table 2.1. Other data are as follows:

Original oil in place = $116.5 \times 10^6$ STB,
Interstitial water saturation = 28.5%,
Bubble point pressure = 3548 psig,
Bubble point oil formation volume factor = 1.450 bbl/STB.

Determine the relative permeability–liquid saturation relationship.

**Table 2.1 Production and Fluid Property Data for a Sand Reservoir**

| Pressure | Instantaneous Gas-Oil Ratio | Oil Formation Volume Factor | Gas Formation Volume Factor | Solution Gas-Oil Ratio | Cumulative Oil Production (MMSTB) | Oil Viscosity | Gas Viscosity |
|---|---|---|---|---|---|---|---|
| 3548 | 770 | 1.450 | 0.000815 | 770 | 0 | 0.49 | 0.01612 |
| 3448 | 850 | 1.443 | 0.000840 | 752 | 0.476 | 0.54 | 0.01682 |
| 3303 | 920 | 1.432 | 0.000875 | 725 | 1.743 | 0.60 | 0.01765 |
| 3153 | 990 | 1.420 | 0.000910 | 695 | 2.818 | 0.67 | 0.01821 |
| 2938 | 1000 | 1.403 | 0.000970 | 657 | 4.632 | 0.73 | 0.01901 |
| 2813 | 1020 | 1.393 | 0.001010 | 632 | 6.030 | 0.80 | 0.01975 |
| 2678 | 1180 | 1.382 | 0.001062 | 608 | 7.360 | 0.89 | 0.02099 |
| 2533 | 1420 | 1.371 | 0.001122 | 580 | 8.751 | 0.98 | 0.02248 |
| 2453 | 1510 | 1.364 | 0.001162 | 565 | 9.873 | 1.09 | 0.02396 |
| 2318 | 1660 | 1.354 | 0.001230 | 540 | 11.259 | 1.20 | 0.02500 |

**INPUT LISTING**

# FORTRAN CODING FORM

UU/CC  NAME  DATE

PROGRAM  
ROUTINE

0 = ZERO   1 = ONE   2 = TWO  
Ø = ALPHA O   I = ALPHA I   Z = ALPHA Z

| CARD | STATEMENT NUMBER | Data |
|---|---|---|
| 1 | PROBLEM #8 | LOS ANGELES        CALIFORNIA       P.-137 |
| 2 | 101.-2 | 85I.-45  II.65,0,0,0,0.-.3,5,4,8 |
| 3-1 | 3,5,4,8 | .7,2,0.-1.-4,5,0.-.0,0,8,1,5.-.7,7,0.-  0.-  .1,9.-  .0,1,6,1,1,2 |
| 3-2 | 3,4,4,8 | .8,5,0.-1.-4,4,3.-.0,0,8,4,0.-.7,5,1,2.-  4,7,1,6,0,0,0.-  .5,4.-  .0,1,6,8,1,2 |
| 3-3 | 3,3,0,3 | .9,2,0.-1.-4,3,2.-.0,0,8,7,5.-.7,2,1,5.-  1,7,4,3,0,0,0.-  .6,0.-  .0,1,7,6,5 |
| 3-4 | 3,1,5,3 | .9,9,0.-1.-4,2,0.-.0,0,9,1,0.-.6,9,5.-  2,8,1,8,0,0,0.-  .6,7.-  .0,1,8,2,1 |
| 3-5 | 2,9,3,8 | 1,0,0,0.-1.-4,0,3.-.0,0,9,7,0.-.6,5,7.-  9,6,3,2,0,0,0.-  .7,3.-  .0,1,9,0,1 |
| 3-6 | 2,8,1,3 | 1,0,2,0.-1.-3,9,3.-.0,0,1,0,1,0.-.6,3,2.-  6,0,3,0,0,0,0.-  .8,0.-  .0,1,9,7,5 |
| 3-7 | 2,6,7,8 | 1,1,8,0.-1.-3,8,2.-.0,0,1,0,6,2.-.6,0,8.-  7,3,4,0,0,0,0.-  .8,9.-  .0,2,1,0,9,9 |
| 3-8 | 2,5,3,3 | 1,4,2,0.-1.-3,7,1.-.0,0,1,1,2,2.-.5,8,0.-  8,7,5,1,0,0,0.-  .9,8.-  .0,2,1,2,4,8 |
| 3-9 | 2,4,5,3 | 1,5,1,0.-1.-3,6,4.-.0,0,1,1,6,2.-.5,6,5.-  9,8,7,3,0,0,0.-  1..0,9.-  .0,2,3,9,6 |
| 3-10 | 2,3,1,8 | 1,6,6,0.-1.-3,5,4.-.0,0,1,2,3,0.-.5,4,0.-  1,1,2,5,9,0,0,0.-  1..2,0.-  .0,2,5,0,0 |

## SOURCE LISTING

```
      THIS PROGRAM IS DESIGNED FOR ESTIMATING GAS-OIL RELATIVE
      PERMEABILITY VALUES FROM ACTUAL PRODUCTION HISTORY FOR A
      SOLUTION GAS DRIVE RESERVOIR

      DIMENSION C1(25),C2(25),C3(25),C4(25),C5(25),C6(25),C7(25),
     *C8(25),C9(25),C10(25),C11(25),C12(25),C13(25),C14(25),C15(25),
     *C16(25),C17(25),C18(25),A(20)
      C1=RESERVOIR PRESSURE, PSIA
      C2=INST.GOR, SCF/STB
      C3=OIL FVF, RES.BBL/STB
      C4=GAS FVF, RES.BBL/SCF
      C5=SOLN.GOR, SCF/STB
      C7=OIL VISCOSITY, CP
      C18=GAS VISCOSITY, CP


      READ(5,1)(A(I),I=1,20)
    1 FORMAT(20A4)
      WRITE(6,12)
   12 FORMAT(//,3X,'GAS-OIL RELATIVE PERMEABILITY FROM PRODUCTION'
     *,' DATA',/,3X,'SOLUTION GAS DRIVE RESERVOIR')
      WRITE(6,2)(A(I),I=1,20)
    2 FORMAT(//,3X,20A4)
      READ(5,3)N,SWI,BOB,OOIP,BPP
    3 FORMAT(I2,2F5.3,F10.0,F6.0)
      N=NUMBER OF DATA POINTS
      SWI=INITIAL WATER SATURATION
      BOB=FORMATION VOLUME FACTOR FOR OIL AT BUBBLE POINT
      OOIP=ORIGINAL OIL IN PLACE, BBL
      BBP=BUBBLE POINT PRESSURE, PSI
      WRITE(6,4)OOIP
    4 FORMAT(/,3X,'ORIGINAL OIL IN PLACE',F22.0,1X,'STB')
      WRITE(6,5)SWI
    5 FORMAT(/,3X,'INITIAL WATER SATURATION (AVG)',F8.3)
      WRITE(6,6)BPP
    6 FORMAT(/,3X,'BUBBLE POINT PRESSURE',F14.0,' PSI')
      DO 10 I=1,N
      READ(5,7)C1(I),C2(I),C3(I),C4(I),C5(I),C6(I),C7(I),C18(I)
    7 FORMAT(2F6.0,F5.3,F8.6,F6.0,F10.0,F5.2,F10.6)
      IF(C1(I).GE.BPP)C2(I)=C5(I)
      C9(I)=C4(I)*C18(I)/(C3(I)*C7(I))
      C10(I)=C2(I)-C5(I)
      C11(I)=C10(I)*C9(I)
      C12(I)=OOIP-C6(I)
      C13(I)=C12(I)/OOIP
      C14(I)=C3(I)/C3(1)
      C15(I)=C13(I)*C14(I)
      C16(I)=(1.-SWI)*C15(I)
      C17(I)=SWI+C16(I)
   10 CONTINUE
      WRITE(6,8)
    8 FORMAT(//,3X,'PRESSURE',3X,'BP PRESS',3X,'INST.GOR',3X,
     *'SOL.GOR',3X,'LIQ.SAT',4X,'KG/KO')
      DO 20 I=1,N
      WRITE(6,9)C1(I),BPP,C2(I),C5(I),C17(I),C11(I)
    9 FORMAT(F9.0,F10.0,F12.0,F10.0,F12.5,F11.5)
   20 CONTINUE
      STOP
      END
//GO.SYSIN DD *
PROBLEM #8      LOS ANGELES     CALIFORNIA      P 37
10  .2851.450116500000. 3548.
3548.   770. 1.450 .000815    770. 0.           .49        .01612
3448.   850. 1.443 .000840    752. 476000.      .54        .01682
3303.   920. 1.432 .000875    725. 174300C.     .60        .01765
3153.   990. 1.420 .000910    695. 2818000.     .67        .01821
2938.  1000.1.403 .000970     657. 4632000.     .73        .01901
2813.  1020.1.393 .001010     632. 6C30000.     .80        .01975
2678.  1180.1.382 .001062     608. 7360000.     .89        .02099
2533.  1420.1.371 .001122     580. 8751000.     .98        .02248
2453.  1510.1.364 .001162     565. 9873000.    1.09        .02396
2318.  1660.1.354 .001230     540. 11259000.   1.20        .02500
/*
```

## OUTPUT LISTING

```
GAS-OIL RELATIVE PERMEABILITY FROM PRODUCTION DATA
SOLUTION GAS DRIVE RESERVOIR

PROBLEM #8      LOS ANGELES     CALIFORNIA      P 37

ORIGINAL OIL IN PLACE                116500000. STB

INITIAL WATER SATURATION (AVG)     0.285

BUBBLE POINT PRESSURE              3548. PSI
```

| PRESSURE | BP PRESS | INST.GOR | SOL.GOR | LIQ.SAT | KG/KO |
|---|---|---|---|---|---|
| 3548. | 3548. | 770. | 770. | 1.00000 | 0.0 |
| 3448. | 3548. | 850. | 752. | 0.99364 | 0.00178 |
| 3303. | 3548. | 920. | 725. | 0.98056 | 0.00351 |
| 3153. | 3548. | 990. | 695. | 0.96827 | 0.00514 |
| 2938. | 3548. | 1000. | 657. | 0.94932 | 0.00618 |
| 2813. | 3548. | 1020. | 632. | 0.93634 | 0.00695 |
| 2678. | 3548. | 1180. | 608. | 0.92342 | 0.01037 |
| 2533. | 3548. | 1420. | 580. | 0.91026 | 0.01577 |
| 2453. | 3548. | 1510. | 565. | 0.90059 | 0.01770 |
| 2318. | 3548. | 1660. | 540. | 0.88814 | 0.02120 |

## BIBLIOGRAPHY

Guerrero, E. T.: *Practical Reservoir Engineering,* The Petroleum Publishing Co., Tulsa (1968).

Smith, Charles Robert: *Mechanics of Secondary Oil Recovery,* Reinhold Publishing Corporation, New York (1966).

# 3    CONVENTIONAL WELL LOG ANALYSIS

**PURPOSE**

The program is designed to provide the values of water saturation, porosity, hydrocarbon index, and movable oil index for each subinterval analyzed. In addition, the program provides average values of porosity, water saturation (Archie method, 1942), and net pay thickness for the total interval based on user-selected cut-off parameters. Formation factors are computed based on the type of lithology selected by the user.

Formation evaluation for the determination of commercial oil and gas potential requires knowledge of porosity, pay zone thickness (net pay), and fluid saturation distribution. It is also often helpful to determine some quantitative index indicating the presence of hydrocarbon and whether such hydrocarbon is producible in commercial quantities (movable oil index). Three types of porosity logs (sonic, density, and neutron) are available, and while all porosity logs are primarily responsive to porosity, other formation characteristics also influence the measurements. Each device responds differently to the effects of rock lithology and type and quantity of the fluids (water, oil, gas) in the pore space. Thus, a combination of two or three porosity logs may often yield a better understanding of porosity, lithology, and pore geometry.

In clean, highly porous, and permeable formation, sonic, formation density, and neutron tools all give sufficiently accurate porosity values. For shaley sand, the formation density log appears to provide the best value of porosity.

Of the formation parameters obtained directly from logs, resistivity is of particular importance. Resistivity measurements, with porosity and water resistivity, are used to obtain water saturation. Water saturation values may be computed by various methods such as (1) the Archie method, (2) the ratio method, (3) the shaley sand method, (4) the two-porosity method, and (5) the three-porosity method.

A clean sand made up of large-to-medium-sized grain and containing hydrocarbon in the intergranular porosity shows a high degree of resistivity in comparison to nearby water sand. For such sand, the Archie method provides a reasonable estimate of water saturation. The ratio method of water saturation determination provides a verification and an improvement to the values generated by the Archie method.

The presence of argillaceous material (shale) in a sandstone formation lowers the formation's resistivity ($R_T$) and, hence, low resistivity contrast. Hydrocarbon zones, under such conditions, often calculate high water saturation by the Archie and ratio methods. The shaley material in the sand lowers the resistivity of the formation because of adsorbed water on the clay particles. Therefore, for shaley sand, modifications must be made to the Archie formula to obtain a more reliable estimate of water saturation value. The shaley sand analysis presented in this program is based on Schlumberger's (1975) shaley sand nomogram equation. One must have two-porosity log data (density and neutron) available along with gamma ray and electric resistivity log data for utilizing this method of computation.

Silts in the formation also lead to low resistivity in oil and gas sands. The extremely small grain size of silt leads to high irreducible water saturation. A clay-free silty sand with high water saturation nonetheless may produce clean oil or gas.

A low resistivity formation containing shale, silt, and hydrocarbon is best analyzed when all three porosity (density, neutron, and sonic) values are used to supplement the resistivity measurement. Each of the porosity log measurement devices is sensitive to porosity as well as presence of hydrocarbon and clay/silt. An important by-product of the method is an indication of abnormally pressured sand ($C_p$).

However, often all three porosity logs may not become available for such formation. In such a situation, two-porosity (density and neutron) logs may be used for the analysis for water saturation with somewhat decreased reliablility.

## METHOD

$F = 1.45/\phi^{1.54}$ for average sands

$F = 1.65/\phi^{1.33}$ for shaley sands

$F = 1.45/\phi^{1.70}$ for calcareous sands

$F = 0.85/\phi^{2.14}$ for carbonates

$S_{WA} = \sqrt{FR_W/R_T}$

$S_{WR} = \left[\dfrac{R_{xo}/R_T}{R_{mf}/R_W}\right]^{5/8}$

If $S_{WR} > S_{WA}$, then $S_{WR} = S_{WA}\left(\dfrac{S_{WA}}{S_{WR}}\right)^{0.25}$

$$\phi_S = \phi_e + q\phi_z + (C_p - 1)\phi_e S_{gxo}$$

$$\phi_D = \phi_e + 0.5\ \phi_e S_{gxo}$$

$$\phi_N = \phi_e + q\phi_z - 0.7\ \phi_e S_{gxo}$$

$$F_Z = 0.62/\phi_z^{2.15}$$

$$S_{WTH} \text{ or } S_{WTW} = \frac{\sqrt{\dfrac{F_Z R_W}{R_T} + \left[\dfrac{q(R_c - R_W)}{2R_c}\right]^2} - \dfrac{q(R_c + R_W)}{2R_c}}{1 - q}$$

$$IGR = \frac{GR - GR_{Min}}{GR_{Max} - GR_{Min}}$$

$$V_{SH} = 0.08336(2^{3.7 IGR} - 1)$$

$$\phi_{NC} = \phi_N - 0.667\phi_{Clay}V_{SH}$$

$$\phi_{DC} = \phi_D - 0.2889\phi_{Clay}V_{SH}$$

$$\phi_{ND} = \sqrt{\frac{\phi_{NC}^2 + \phi_{DC}^2}{2}}$$

$$S_{WSH} = -\frac{\left(\dfrac{-V_{SH}}{R_{SH}}\right) + \sqrt{\left(\dfrac{V_{SH}}{R_{SH}}\right)^2 + \dfrac{\phi_{ND}^2}{[0.2R_W(1 - V_{SH})R_T]}}}{\dfrac{\phi_{ND}^2}{0.4R_W(1 - V_{SH})}}$$

where

$F$ = Formation factor;
$\phi$ = Porosity, fraction;
$S_{WA}$ = Water saturation by Archie method, fraction;
$S_{WR}$ = Water saturation by ratio method, fraction;
$\phi_s$ = Sonic porosity, fraction;
$\phi_e$ = Effective porosity, fraction;
$\phi_z$ = Total porosity of the sand matrix, including pore fluids and clay;
$q$ = Shale/silt fraction in the rock;
$C_p$ = Compaction factor;
$S_{gxo}$ = Residual hydrocarbon saturation in the zone close to the borehole;
$\phi_N$ = Neutron porosity, fraction;
$\phi_D$ = Density porosity, fraction;
$S_{WTH}$ = Water saturation by three-porosity method. fraction;
$S_{WTW}$ = Water saturation by two-porosity method, fraction;

$V_{SH}$ = Shale fraction;

$\phi_{NC}$ = Corrected neutron porosity, fraction;

$\phi_{DC}$ = Corrected density porosity, fraction;

$\phi_{ND}$ = Average neutron-density porosity, fraction;

$S_{WSH}$ = Water saturation by shaley sand method, fraction;

$R_{SH}$ = Resistivity of shale;

$R_W$ = Resistivity of formation water;

$R_T$ = Formation resistivity;

$\phi_{\text{clay}}$ = Porosity of shale, fraction.

## PROGRAM DESCRIPTION

The program consists of a FORTRAN main program and can handle up to 100 subinterval analysis. However, no restriction is placed on the number of intervals that can be analyzed, but no interval should be broken down to have more than 100 subintervals. Users have the option to select from 4 different lithology types. Water saturation by the Archie method and the ratio method is computed for each subinterval. However, other methods (such as shaley sand, two porosity, and three porosity) of water saturation computations are carried out at the user's option. The user must also provide the program with the type of electric log combination data being used for the evaluation—e.g., dual induction laterolog or dual laterolog with a microfocused log. Various cut-off values such as water saturation (Archie equation), density porosity, and shale fraction are also provided as input parameters for net pay, average water saturation, and porosity computation.

### Input Format

| Card | Format | Content |
|------|--------|---------|
| 1 | 20A4 | X2(I) |
| 2 | 6I5 | INTN, ISHLY, ITWPOR, ITHPOR, LTYPE, TRL |
| 3 | 8F8.2 | RMAXT, DMAXT, RMF, TRMF, SMT, GG, GRM, GRL |
| 4–INTN | 2F10.1, F5.3, F10.2, F5.0, F7.0, 3F3.2, F5.1, F3.2, I3 | TZ, BZ, RW, SP, TRW, SLT, SWC, SHCO, OCT, RSH, ONCLAY, N1 |
| 5–N1 (Required only if TRL = 1) | 2F6.0, F5.0, 3F5.1, 3F5.3 | T(J), B(J), GR(J), RILD(J), RILM(J), RLL8(J), OD(J), ON(J), OS(J) |
| 6–N1 (Required only if TRL = 2) | 2F6.0, F5.0, 3F5.1, 3F5.3 | T(J), B(J), GR(J), RLLD(J), RLLS(J), SFLM(J), OD(J), ON(J), OS(J) |

**Description of Input Parameters**

X2(I) = Up to 80 alphanumeric characters for project identification;

INTN = Number of intervals to be analyzed;

ISHLY = Shaley sand analysis option: ISHLY = 1 indicates shaley sand analysis is desired;

ITWPOR = Two-porosity analysis option: ITWPOR = 0 indicates that two-porosity analysis is not desired;

ITHPOR = Three-porosity analysis option: ITHPOR = 0 indicates that three-porosity analysis is not desired;

LTYPE = Lithology type option: LTYPE = 1 indicates clean sand, LTYPE = 2 indicates shaley sand, LTYPE = 3 indicates calcareous sand, and LTYPE = 4 indicates carbonates;

TRL = Option for type of electric log desired: TRL = 1 indicates dual induction laterolog combination values to be used, TRL = 2 indicates dual laterolog microfocused combination values to be used;

RMAXT = Maximum recorded temperature, °F;

DMAXT = Depth of maximum recorded temperature, ft;

RMF = Resistivity of mud filtrate;

TRMF = Temperature at which mud filtrate resistivity was measured;

SMT = Surface mean temperature;

GG = Geothermal gradient (deg/ft). May set equal to zero, in which case it will be internally computed;

GRM = Maximum gamma ray log reading;

GRL = Minimum gamma ray log reading;

TZ = Top of the interval, ft;

BZ = Bottom of the interval, ft;

RW = Formation water resistivity: if not provided by the user, it will be computed from the SP log data internally, provided water salinity is not known as well;

SP = Corrected SP reading of the zone;

TRW = Temperature at which RW was measured;

SLT = Salinity of the formation water: if RW is not known but salinity is known, RW will be computed using salinity value;

SWC = Water saturation cut-off value;

SHCO = Shale fraction cut-off value;

OCT = Density porosity cut-off value;

RSH = Resistivity of shale: if not known, input 0. Default is 10 times the formation water resistivity;

ONCLAY = Clay porosity: if unknown, use 0. Default value is 30%;

N1 = Number of subinterval within the zone or interval at which computations are to be performed (not to exceed 100);

T(J) = Top of the subinterval, ft;

B(J) = Bottom of the subinterval, ft;

GR(J) = Gamma ray reading;

RILD(J) = Deep induction resistivity value;

RILM(J) = Medium induction resistivity value;

RLL8(J) = Laterolog 8 resistivity value;
OD(J) = Density porosity value;
ON(J) = Neutron porosity value;
OS(J) = Sonic porosity value;
RLLD(J) = Deep laterolog resistivity value;
RLLS(J) = Shallow laterolog resistivity value;
SFML(J) = Microfocused log resistivity value.

## Description of Output Parameters

TOP = Top of the subinterval analyzed, ft;
BOT = Bottom of the subinterval analyzed, ft;
POR = Density porosity, fraction;
SWA = Water saturation by Archie method, fraction;
SWR = Water saturation by ratio method, fraction;
MOI = Movable oil index; less than 0.7 indicates oil is movable, while values greater than 0.7 oil is not movable;
SWS = Water saturation by shaley sand method, fraction;
SW3 = Water saturation by three-porosity method, fraction;
SW2 = Water saturation by two-porosity method, fraction;
Q = Shale content, fraction;
CP = Compaction correction factor;
RW = Formation water resistivity;
SAL = Formation water salinity, ppm;
HYI = Hydrocarbon index, 1 or O: 1 indicates presence of hydrocarbon and O indicates absence of hydrocarbon;
BVF = Bulk volume fraction water.

## EXAMPLE PROBLEM

The input data listing shows the values obtained from a conventional log suite run in a well. Evaluate two main zones from 10500 to 10512 ft and 10533 to 10550 ft. Provide zonal analysis on a 2-ft subinterval basis. Since both of these zones appear to contain shale/silt, determine the water saturation by the shaley sand, three-porosity, and two-porosity methods for comparison. A dual induction–laterolog combination resistivity device along with an SP device were run in the well. Gamma ray, density, and neutron devices as well as sonic tools were run.

# INPUT LISTING

0 = ZERO    1 = ONE    2 = TWO
Ø = ALPHA O    I = ALPHA I    Z = ALPHA Z

| STATEMENT NUMBER | | | | | | 6 7 8 9 10 ... 67 |
|---|---|---|---|---|---|---|
| | | | | | | PROBLEM NO. 118 UINKNOWN MIE LII X.Y. FORMATION |
| | | | | | | 2 |
| 12:50: | 12:0:00: | 1:0:16 | 1:15: | 1:6:0: | 0:0:0: | 4:0:0: |
| 10:5:00 | 12:0:00: | 1:0:8: | 1:2:0:0: | 6:8: | 1:0:1:5:0- 1:5:0 | 1:0: |
| 10:5:00 | 1:0:5:0:2 | 1:2:1:0 | 2:0:0 | 2:2:0 | 2:5:0 1:0:1:4:3:0 1:0:8:0 0:1:5:5 | 1:0: |
| 10:5:01:2 | 1:0:5:0:4 | 2:0:8:0 | 2:5: | 2:1:6:0 | 3:0:0 1:0:6:8:0 0:9:0 0:8:0 | |
| 10:5:04 | 1:0:5:0:6 | 2:0:2:0 | 2:8:0 | 2:7:0 | 3:2:0 1:0:7:0 0:9:3 1:0:8:5 | |
| 10:5:06 | 1:0:5:0:8 | 1:2:0:1:0 | 3:0:0 | 3:2:0 | 3:5:0 1:0:9 0:8:1 1:2:0 | |
| 10:5:08 | 1:0:5:0:9 | 1:9:7:0 | 4:0:0 | 4:2:0 | 4:5:0 1:1:0 0:8:5 1:2:5 | |
| 10:5:09 | 1:0:5:1:0 | 1:9:5:0 | 4:0:0 | 4:4:0 | 4:8:0 1:5:0 0:8:2 1:6:0 | |
| 10:5:10 | 1:0:5:1:2 | 1:9:3:0 | 1:5:0:0 | 5:3:0 | 5:5:0 1:4:0 0:8:5 1:5:5 | |
| 10:5: | 5:3:0 | 1:6:5:0 | 1:0:6 | 1:2:5:0 | 6:8:0 0:0:1:5:0 3:0:0:1:6 0:0:1:3:0 | |
| 10:5:33 | 1:0:5:3:4 | 2:0:2:0 | 2:9:0 | 2:9:5 | 3:1:5:0 1:0:7:0 0:7:1:5 1:0:8:5 | |
| 10:5:34 | 1:0:5:3:7 | 1:8:7:0 | 4:7:0 | 4:9:0 | 5:5:0 1:0:1:6 0:8:0 1:1:2:5 | |
| 10:5:37 | 1:0:5:4:0 | 1:8:8:0 | 6:5:0 | 6:1:8:0 | 1:5:0 1:2:3:0 1:8:0 1:4:5 | |
| 10:5:40 | 1:6:5:4:3 | 1:9:2:0 | 5:0:0 | 5:5:0 | 5:9:0 1:1:3:5 0:7:5 1:4:5 | |
| 10:5:43 | 1:0:5:4:5 | 1:9:5:0 | 4:3:0 | 4:6:0 | 4:9:0 1:0:1:3 0:7:5 1:5:5 | |
| 10:5:45 | 1:0:5:4:6 | 1:9:7:0 | 4:5:0 | 4:9:0 | 4:9:0 1:4:0 0:8:0 1:6:0 | |
| 10:5:46 | 1:0:5:4:7 | 1:9:6:0 | 4:2:0 | 4:2:0 | 4:8:0 1:1:5 0:8:3 1:3:0 | |
| 10:5:47 | 1:6:5:5:6 | 1:9:5:0 | 4:1:0 | 4:1:0 | 4:4:0 1:5:0 0:7:2 1:7:0 | |

## SOURCE LISTING

```
C       THIS PROGRAM IS DESIGNED TO PERFORM CONVENTIONAL LOG ANALYSIS AND
C       PROVIDES WATER SATURATION ESTIMATES BASED ON VARIOUS CALCULATION
C       PROCEEDURES SUCH AS ARCHI METHOD, RATIO METHOD, TWO POROSITY AND
C       THREE POROSITY METHOD, SHALY SAND METHOD.  IT ALSO ATTEMPTS TO
C       PROVIDE AN ESTIMATE OF HYDROCARBON INDEX AND MOVABLE OIL INDEX.
C
C       INTN=NO. OF ZONES TO BE EVALUATED
C       ISHLY=SHALEY SAND ANALYSIS FLAG; 1=SHALEY SAND ANALYSIS IS REQ'D
C       ITWPOR=TWO PORCSITY ANALYSIS INDEX, 0=2 POROSITY ANALYSIS IS NOT
C       DESIRED
C       ITHPOR=THREE POROSITY ANALYSIS INDEX, 0=3 POROSITY ANALYSIS IS NOT
C       DESIRED
C       LTYPE=LITHOLOGY TYPE, 1=CLEAN SAND
C                             2=SHALEY SAND
C                             3=CALCARED SAND
C                             4=CARBONATES
C       TRL=TYPE OF ELECTRICAL LOG CCMBINATION USED, 1=DUAL INDUCTION LATERO
C       LOG, 2=DUAL LATEROLOG AND MICROFOCOUSED LOG
C       RMAXT=MAX RECORDED TEMP
C       DMAXT=MAX TEMP DEPTH
C       RMF=FILTRATE RESISTIVITY TEMPERATURE
C       SMT=SURFACE MEAN TEMP
C       GG=GEOTHERMAL TEMP GRAD
C       GRM=MAX GR READING
C       GRL=.IN GR READING
C       TZ=TOP OF ZONE
C       BZ=BOTTOM OF ZONE
C       RW=FORMATION OF WATER RESISTIVITY
C       SP=SP READING
C       TRW=FORMATION WATER SALINITY
C       SWC=WATER SATURATICN CUT-OFF
C       SHCO=SHALE CONTENT CUT-OFF
C       OCT=DENSITY POROSITY CUT-CFF
C       RSH=SHALE RESISTIVITY
C       ONCLAY=CLAY POROSITY
C       GR=GAMMA-LOG READING
C       RILD=DEEP INDUCTION LOG READING
C       RILM=MEDIUM INDUCTION LOG READING
C       RLL8=LATEROLOG 8 READING
C       ON=NEUTRON POROSITY READING
C       OS=SONIC POROSITY READING
C
        DIMENSION X2(20),T(100),B(100),RILD(100),RLL8(100),RILM(100),
       *RT(100),RLLD(100),RLLS(100),SFLM(100),RTRILD(100),OD(100),ON(100)
       *,OS(100),SWA(100),GR(100),VSH(100),HYI(100)
        INTEGER TRL
        READ(5,100)(X2(I),I=1,20)
100     FORMAT(20A4)
        READ(5,101)INTN,ISHLY,ITWPOR,ITHPOR,LTYPE,TRL
101     FORMAT(6I5)
        READ(5,102)RMAXT,DMAXT,RMF,TRMF,SMT,GG,GRM,GRL
102     FORMAT(8F8.2)
        WRITE(6,103)
103     FORMAT(///,30X,' LOG INTERPRETATION')
        WRITE(6,104)(X2(I),I=1,20)
104     FORMAT(//,3X,20A4)
        RMF75=RMF*(TRMF+7.)*.012195
        IF(GG.EQ.0.)GG=(RMAXT-SMT)/DMAXT
        DO 1000 I=1,INTN
        READ(5,105)TZ,BZ,RW,SP,TRW,SLT,SWC,SHCO,OCT,RSH,ONCLAY,N1
        IF(RW.EQ.0..AND.SLT.EQ.0)WRITE(6,111)
111     FORMAT(/,2X,'CATALOGED VALUE OF RW IS NOT AVAILABLE.  RW VALUE'
       *,' IS CALCULATED FROM SP MEASUREMENTS, PLEASE CHECK SP VALJES',
       *' IN CASE OF DOUBT')
        IF(RW.GT.0.)WRITE(6,112)RW,TRW
112     FORMAT(/,3X,'CATALOGED VALUE OF RW = ',F6.4,F7.0,1X,'F')
        IF(RW.EQ.0..AND.SLT.GT.0.)WRITE(6,113)SLT
113     FORMAT(/,3X,'RW IS CALCULATED FROM SALINITY',3X,F6.0,1X,'PPM')
105     FORMAT(2F10.1,F5.3,F10.2,F5.0,F7.0,3F3.2,F5.1,F3.2,I3)
        GI=BZ-TZ
        ITZ=TZ
        IBZ=BZ
        IGI=GI
        WRITE(6,106)I,ITZ,IBZ,IGI
106     FORMAT(//,14X,'INTERVAL ',I3,' ANALYZED',3X,I5,'FT',' -',I6,
       *'FT',3X,'GROSS INTERVAL OF',I6,'FT')
        WRITE(6,107)
107     FORMAT(/,3X,'TOP',5X,'BOT',5X,'POR',2X,'SWA',3X,'SWR',3X,'MOI',
       *3X,'SWS',3X,'SW3',3X,'SW2',3X,'Q',5X,'CP',3X,'RW',5X,'SAL',
       *4X,'HYI',3X,'BVF')
```

```
         IF(RW.GT.0.)RW75=RW*(TRW+7)*0.012195
         SPOR=0.
         SXNET=0.
         SSWA=0.
         DO 1001 J=1,N1
         IF(TRL.EQ.1)READ(5,108)T(J),B(J),GR(J),RILD(J),RILM(J),RLL8(
       *J),OD(J),ON(J),OS(J)
         IF(TRL.EQ.2)READ(5,108)T(J),B(J),GR(J),RLLD(J),RLLS(J),SFLM(
       *J),OD(J),ON(J),OS(J)
  108 FORMAT(2F6.0,F5.0,3F5.1,3F5.3)
         D1=(B(J)-T(J))*.5+T(J)
         T1=D1*GG+SMT
         RMF1=RMF*(TRMF+7)/(T1+7)
         IF(RW.GT.0.)GO TO 1002
         IF(SLT.GT.0.)GO TO 1003
         XK=60.+.133*T1
         RMFRWE=10.**(-SP/XK)
         IF(RMF75.GT.0.1)RMF1=.85*RMF1
         IF(RMF75.LE.0.1)RMF1=(146.*RMF1-5.)/(337.*RMF+77.)
         RWE=RMF1/RMFRWE
         IF(RWE.LT.0.12)RW75=(77.*RWE+5.)/(146.-337.*RWE)
         IF(RWE.GE.0.12)RW75=-(0.58-10.**(0.69*RWE-0.24))
         GO TO 1002
 1003 CONTINUE
         RW75=10.**(3.562-.955*ALOG10(SLT))+.0123
 1002 CONTINUE
         RW=RW75*82./(T1+7.)
         IF(SLT.EQ.0.)SLT=10.**((3.562-ALOG10(RW75-0.0123))/.955)
         IF(TRL.EQ.2)GO TO 1004
         IF((RLL8(J)/RILD(J)).LT.2.5)GO TO 1005
         IF((RILM(J)/RILD(J)).LT.1.)GO TO 1005
         RTRILD(J)=1.-(((RILM(J)/RILD(J)-1.1)**2)*((.4*RLL8(J)/RILD(J)
       *)*(RLL8(J)/RILD(J)+3))-1.)/(RLL8(J)/RILD(J)-1.)**2
         RT(J)=RILD(J)*RTRILD(J)
         IF((RT(J)/RILD(J)).LT.0.4)GO TO 1005
         GO TO 1006
 1005 CONTINUE
         RT(J)=RILD(J)
         GO TO 1006
 1004 CONTINUE
         RT(J)=((RLLD(J)-.29*SFLM(J))/0.71)*(RLLD(J)/RLLS(J))**.1
         IF(RLLD(J).LT.RLLS(J))GO TO 1007
         IF(RLLD(J).LT.SFLM(J))GO TO 1007
         GO TO 1006
 1007 RT(J)=RLLD(J)
 1006 CONTINUE
         IF(RSH.EQ.0.)RSH=10.*RW
         IF(ONCLAY.EQ.0.)ONCLAY=0.3
         SWSCH=0.
         SWTH=0.
         SWTW=0.
         IF(LTYPE.EQ.1)F=1.45/(OD(J)**1.54)
         IF(LTYPE.EQ.2)F=1.65/(OD(J)**1.33)
         IF(LTYPE.EQ.3)F=1.45/(OD(J)**1.7)
         IF(LTYPE.EQ.4)F=0.85/(OD(J)**2.14)
         SWA(J)=SQRT(F*RW/RT(J))
         IF(TRL.EQ.1)SWR=((RLL8(J)/RT(J))/(RMF1/RW))**.625
         IF(TRL.EQ.2)SWR=((SFLM(J)/RT(J))/(RMF1/RW))**.625
         IF(SWR.GT.(1.1*SWA(J)))SWR=SWA(J)*(SWA(J)/SWR)**0.25
         RWA=RT(J)/F
         IF(RWA.GT.(3.*RW))HYI(J)=1.
         IF(RWA.LE.(3.*RW))HYI(J)=0.
         IF(TRL.EQ.1)SWSXO=SQRT((RLL8(J)/RILD(J))/(RMF1/RW))
         IF(TRL.EQ.2)SWSXO=SQRT((SFLM(J)/RILD(J))/(RMF1/RW))
         IF(ISHLY.EQ.1) GO TO 1020
         GO TO 1030
 1020 CONTINUE
         XIGR=(GR(J)-GRL)/(GRM-GRL)
         VSH(J)=.08336*(2**(3.7*XIGR)-1.)
         ONC=ON(J)-0.6667*CNCLAY*VSH(J)
         ODC=OD(J)-0.2889*CNCLAY*VSH(J)
         ONOD=SQRT((ONC**2+ODC**2)*.5)
         SWSCH=(-VSH(J)/RSH+SQRT((VSH(J)/RSH)**2+ONOD**2/(0.2*RW*(1.-
       *VSH(J))*RT(J))))/(ONOD**2/(0.4*RW*(1.0-VSH(J))))
 1030 CONTINUE
         IF(ITHPOR.EQ.0) GO TO 1040
         CP=1.
         DO 1100 K=1,100
         OSC=OS(J)
         OE=OD(J)-.5*(OSC-ON(J))/(CP-.3)
         SGXO=2.*OD(J)/OE-2.
         OZ=ON(J)+.7*OE*SGXO
         O=1.-OE/OZ
         IF(Q.GE.0.0) GO TO 1101
         IF(Q.LT.0.0.AND.CP.GT.1.0) CP=CP-.01
         IF(Q.LT.0.0.AND.CP.EQ.1.0)OS(J)=0.58333*OD(J)+.41667*ON(J)
         GO TO 1100
 1101 CONTINUE
```

```
        FZ=.62/(OZ**2.15)
        SWTH=(SQRT((FZ*RW/RT(J))+(Q*(RSH-RW)/(2.*RSH))**2)-(Q*(RSH+RW)
       */(2.*RSH)))/(1.-Q)
        QCHK=.7*(1.-SWTW)
        IF(SGXO.GT.QCHK)CP=CP+.1
        IF(SGXO.LE.QCHK)GO TO 1040
 1100 CONTINUE
 1040 CONTINUE
        IF(ITWPOR.EQ.0)GO TO 2000
        OZ=(ON(J)+1.4*OD(J))/2.4
        DO 1111 K1=1,100
        OE=OD(J)-0.71429*(OZ-ON(J))
        Q=1.-OE/OZ
        SGXO=2.*(OD(J)/OE-1.)
        IF(Q.GE.0.0)GO TO 1113
        OZ=OZ+.1
        GO TO 1111
 1113 CONTINUE
        FZ=0.62/(OZ**2.15)
        SWTW=(SQRT((FZ*RW/RT(J))+(Q*(RSH-RW)/(2.*RSH))**2)-(Q*(RSH+RW)
       */(2.*RSH)))/(1.-Q)
        QCHK=0.7*(1.-SWTW)
        IF(SGXO.LE.QCHK)GO TO 2000
        IF(SGXO.GT.QCHK)OZ=OZ-.01
 1111 CONTINUE
 2000 CONTINUE
        BVF=OD(J)*SWA(J)
        WRITE(6,110)T(J),B(J),OD(J),SWA(J),SWR,SWSXO,SWSCH,SWTH,SWTW,
       *VSH(J),CP,RW,SLT,HYI(J),BVF
        XNET=B(J)-T(J)
        IF(OD(J).LT.OCT)XNET=0.
        IF(SWA(J).GT.SWC)XNET=0.
        IF(VSH(J).GT.SHCO)XNET=0.
        SXNET=SXNET+XNET
        SPOR=SPOR+OD(J)*XNET
        SSWA=SSWA+SWA(J)*XNET
  110 FORMAT(/,2F8.0,7F6.2,F5.2,F6.2,F5.2,F9.0,F6.2,F6.3)
 1001 CONTINUE
        IF(SXNET.GT.0.0)WAVPOR=SPOR/SXNET
        IF(SXNET.LE.0.0)WAVPOR=0.
        IF(SXNET.GT.0.0)WASAT=SSWA/SXNET
        IF(SXNET.LE.0.0)WASAT=0.
        WRITE(6,150)SXNET
  150 FORMAT(/,3X,'NET PAY THICKNESS',F13.1,' FT')
        WRITE(6,151)WAVPOR
  151 FORMAT(/,3X,'AVERAGE POROSITY',F15.2,'FRACTION')
        WRITE(6,152)WASAT
  152 FORMAT(/,3X,'AVERAGE WATER SATURATION',F8.2,' FRACTION')
 1000 CONTINUE
        WRITE(6,120)
        WRITE(6,121)
        WRITE(6,122)
        WRITE(6,123)
  120 FORMAT(///,4X,'HYI = HYDROCARBON INDEX, 0=NO INDICATION,'
       *,' 1=INDICATION OF HYDROCARBON PRESENT')
  121 FORMAT(/,3X,' MOI = MOVEABLE OIL INDEX, LESS THAN OR EQUAL',
       *' TO 0.7 HYDROCARBON IS FLUSHED, HENCE PRODUCIBLE')
  122 FORMAT(/,3X,' SLT = SALINITY IN PPM')
  123 FORMAT(/,4X,'BVI = BULK VOLUME FRACTION WATER')
        STOP
        END
//GO.SYSIN DD *
PROBLEM NO. 18        UNKNOWN WELL        XY FORMATION              P 29
     2     1     1     1     2     1
   250.0012000.00      .06    75.00    60.00    0.00  400.00   10.00
    10500.0    10512.0  .08    200.0    68.     0..50.50.05  0.0.30    7
 10500.10502.  210.  20.0  22.0  25.0  .043  .080  .055
 10502.10504.  208.  25.0  26.0  30.0  .068  .095  .080
 10504.10506.  202.  28.0  29.0  32.0  .070  .093  .085
 10506.10508.  201.  30.0  32.0  35.0  .109  .087  .120
 10508.10509.  197.  40.0  42.0  45.0  .110  .085  .125
 10509.10510.  195.  40.0  44.0  48.0  .150  .082  .160
 10510.10512.  193.  50.0  53.0  55.0  .140  .085  .155
    10533.0    10550.0  .06    225.0    68.     0..50.30.06  0.0.30    8
 10533.10534.  202.  29.0  29.5  35.0  .07   .075  .085
 10534.10537.  187.  47.0  49.   55.0  .106  .080  .125
 10537.10540.  188.  65.0  68.   69.5  .123  .180  .145
 10540.10543.  192.  50.0  55.   59.0  .135  .075  .145
 10543.10545.  195.  43.0  46.   49.0  .150  .073  .155
 10545.10546.  197.  40.0  45.   49.0  .140  .080  .160
 10546.10547.  196.  40.0  42.   48.0  .113  .083  .130
 10547.10550.  195.  40.0  41.   44.0  .158  .082  .170
/*
//
```

## OUTPUT LISTING

LOG INTERPRETATION

PROBLEM NO. 18     UNKNOWN WELL       XY FORMATION                    P 29
CATALOGED VALUE OF RW = 0.0800     68. F

INTERVAL   1 ANALYZED   10500FT - 10512FT   GROSS INTERVAL OF   12FT

| TOP | BOT | POR | SWA | SWR | MOI | SWS | SW3 | SW2 | Q | CP | RW | SAL | HYI | BVF |
|---|---|---|---|---|---|---|---|---|---|---|---|---|---|---|
| 10500. | 10502. | 0.04 | 0.37 | 0.27 | 1.23 | 0.06 | 0.60 | 0.82 | 0.23 | 1.00 | 0.03 | 100627. | 1.00 | 0.016 |
| 10502. | 10504. | 0.07 | 0.25 | 0.16 | 1.21 | 0.05 | 0.38 | 0.16 | 0.22 | 1.00 | 0.03 | 100627. | 1.00 | 0.017 |
| 10504. | 10506. | 0.07 | 0.23 | 0.15 | 1.18 | 0.04 | 0.32 | 0.14 | 0.21 | 1.00 | 0.03 | 100627. | 1.00 | 0.016 |
| 10506. | 10508. | 0.11 | 0.16 | 0.10 | 1.19 | 0.04 | 0.14 | 0.27 | 0.21 | 1.00 | 0.03 | 100627. | 1.00 | 0.018 |
| 10508. | 10509. | 0.11 | 0.14 | 0.08 | 1.17 | 0.03 | 0.11 | 0.24 | 0.20 | 1.10 | 0.03 | 100627. | 1.00 | 0.016 |
| 10509. | 10510. | 0.15 | 0.11 | 0.06 | 1.21 | 0.03 | 0.13 | 0.19 | 0.20 | 1.40 | 0.03 | 100627. | 1.00 | 0.017 |
| 10510. | 10512. | 0.14 | 0.11 | 0.06 | 1.16 | 0.03 | 0.08 | 0.18 | 0.19 | 1.30 | 0.03 | 100627. | 1.00 | 0.015 |

NET PAY THICKNESS          10.0 FT
AVERAGE POROSITY           0.10 FRACTION
AVERAGE WATER SATURATION   0.17 FRACTION
CATALOGED VALUE OF RW = 0.0600     68. F

INTERVAL   2 ANALYZED   10533FT - 10550FT   GROSS INTERVAL OF   17FT

| TOP | BOT | POR | SWA | SWR | MOI | SWS | SW3 | SW2 | Q | CP | RW | SAL | HYI | BVF |
|---|---|---|---|---|---|---|---|---|---|---|---|---|---|---|
| 10533. | 10534. | 0.07 | 0.19 | 0.13 | 1.05 | 0.03 | 0.23 | 0.34 | 0.21 | 1.00 | 0.02 | 146303. | 1.00 | 0.014 |
| 10534. | 10537. | 0.11 | 0.12 | 0.07 | 1.03 | 0.02 | 0.07 | 0.20 | 0.18 | 1.20 | 0.02 | 146303. | 1.00 | 0.012 |
| 10537. | 10540. | 0.12 | 0.09 | 0.05 | 0.99 | 0.02 | 0.11 | -0.13 | 0.19 | 1.00 | 0.02 | 146303. | 1.00 | 0.011 |
| 10540. | 10543. | 0.13 | 0.10 | 0.05 | 1.04 | 0.02 | 0.10 | 0.17 | 0.19 | 1.40 | 0.02 | 146303. | 1.00 | 0.013 |
| 10543. | 10545. | 0.15 | 0.10 | 0.05 | 1.02 | 0.02 | 0.12 | 0.17 | 0.20 | 1.40 | 0.02 | 146303. | 1.00 | 0.014 |
| 10545. | 10546. | 0.14 | 0.10 | 0.06 | 1.06 | 0.02 | 0.10 | 0.18 | 0.20 | 1.50 | 0.02 | 146303. | 1.00 | 0.015 |
| 10546. | 10547. | 0.11 | 0.12 | 0.07 | 1.05 | 0.02 | 0.08 | 0.20 | 0.20 | 1.20 | 0.02 | 146303. | 1.00 | 0.014 |
| 10547. | 10550. | 0.16 | 0.10 | 0.05 | 1.00 | 0.02 | 0.09 | 0.16 | 0.20 | 1.40 | 0.02 | 146303. | 1.00 | 0.015 |

NET PAY THICKNESS          17.0 FT
AVERAGE POROSITY           0.13 FRACTION
AVERAGE WATER SATURATION   0.11 FRACTION

HYI = HYDROCARBON INDEX. 0=NO INDICATION. 1=INDICATION OF HYDROCARBON PRESENT
MOI = MOVEABLE OIL INDEX. LESS THAN OR EQUAL TO 0.7 HYDROCARBON IS FLUSHED. HENCE PRODUCIBLE
SAL = SALINITY IN PPM
BVF = BULK VOLUME FRACTION WATER

## BIBLIOGRAPHY

Archie, G. E.: "The Electrical Resistivity Log as an Aid in Determining Some Reservoir Characteristics," *J. Pet. Tech.*, **5** (1942) 54–62; *Trans.*, AIME (1942) **146**, 54.

Asquith, George B.: *Log Analysis by Microcomputer,* The Petroleum Publishing Company, Tulsa (1980).

"A Guide to Well Site Interpretation for the Gulf Coast," Schlumberger Well Services (1975) 23–27.

*Well Logging,* Reprint Series No. 1, SPE, Dallas (1971).

# 4  OIL IN PLACE AND RECOVERABLE RESERVE BY THE VOLUMETRIC METHOD

**PURPOSE**

This program computes oil in place and primary reserve by the volumetric method for a given reservoir. Two options are available for the volumetric calculation. The computation may be performed using Isopach data. A second option is to consider the reservoir to be circular and of uniform thickness.

Two options are available for the recovery factor calculation: (1) Arps statistical correlation (Arps 1970) and (2) simple statistical observation of residual oil saturation (Craze, Buckley, and Arps 1945; 1956) for the type of drive mechanism.

The volumetric method of determining oil in place and primary reserve is probably the oldest and most commonly used method. It is perhaps also one of the most abused techniques; therefore, extreme caution and care should be exercised when utilizing it.

Although computation of oil in place and reserve is quite simple and easy to understand, it does require a considerable amount of work and good engineering judgment to arrive at various parameters such as porosity, water saturation, formation volume factor, net sand thickness, extent or size of the reservoir, and the recovery factor (RF). Recovery factor is a function of the prevailing driving mechanism, the reservoir, and the reservoir fluid properties.

**METHOD**

$$\text{Oil in place} = \frac{7758V\phi(1 - S_{wi})}{B_{oi}}$$

$$\text{Recoverable reserve} = \frac{7758V\phi(1 - S_{wi})}{B_{oi}} \times RF$$

$$V = \frac{h_1}{3}(A_1 + A_2 + \sqrt{A_1 A_2}) + \ldots + \frac{h_n}{3}(A_n)$$

For Isopach data option

$$V = A_o h_o$$

For uniformly thick circular reservoir assumption options

### Depletion Drive

Arps correlation:

$$R_F = 0.41815 \left( \frac{\phi(1 - S_{wi})}{B_{ob}} \right)^{+0.1611} \left( \frac{k}{\mu_{ob}} \right)^{+0.0979} (S_{wi})^{+0.3722} \left( \frac{p_b}{p_a} \right)^{+0.1741}$$

Residual oil saturation correlation:

$$R_F = 1 - \frac{B_{oi}(1 - S_{wi} - S_{ga})}{B_{oa}(1 - S_{wi})}$$

### Water Drive

Arps correlation:

$$R_F = 0.54898 \left( \frac{\phi(1 - S_{wi})}{B_o} \right)^{+0.0422} \left( \frac{k \mu_{wi}}{\mu_{oi}} \right)^{+0.0770} (S_{wi})^{-0.1903} \left( \frac{p_i}{p_a} \right)^{-0.2159}$$

Residual oil saturation correlation

$$R_F = \frac{1 - S_{wi} - S_{or}}{1 - S_{wi}}$$

Craze, Buckley, and Arps residual oil saturation correlation:

| Reservoir Oil Viscosity (cp) | $S_{or}$ (percent) |
|---|---|
| 0.2 | 30 |
| 0.5 | 32 |
| 1.0 | 34.5 |
| 2.0 | 37 |

| | |
|---|---|
| 5.0 | 40.5 |
| 10.0 | 43.5 |
| 20.0 | 64.5 |

| Average Reservoir Permeability (md) | Deviation of $S_{or}$ from above trend (percent) |
|---|---|
| 50 | +12 |
| 100 | +9 |
| 200 | +6 |
| 500 | +2 |
| 1000 | −1 |
| 2000 | −4.5 |
| 5000 | −8.5 |

For these equations,

$V$ = Net reservoir sand volume, acre-ft;
$\phi$ = Average reservoir porosity, fraction;
$S_{wi}$ = Average initial water saturation, fraction;
$B_{oi}$ = Initial oil formation volume factor, bbl/STB;
$RF$ = Recovery factor, fraction;
$h_1, h_2 \ldots hn - 1$ = Contour intervals, ft;
$A_1, A_2, \ldots A_n$ = Area enclosed by successive contours, AC (acre);
$h_n$ = Contour interval between $A_n$ contour and zero contour, ft;
$A_o$ = A real extent of circular reservoir, AC;
$h_o$ = Uniform net thickness of the circular reservoir, ft;
$B_{ob}$ = Oil formation volume factor at bubble point, bbl/STB;
$k$ = Reservoir (absolute) permeability, md;
$\mu_{ob}$ = Reservoir oil viscosity at bubble point, cp;
$p_b$ = Bubble point pressure, psi;
$p_a$ = Abandonment pressure, psi;
$B_{oa}$ = Oil formation volume factor at abandonment pressure;
$S_{ga}$ = Gas saturation at abandonment condition, fraction (default value 0.304);
$\mu_{wi}$ = Water viscosity at initial condition, cp;
$\mu_{oi}$ = Oil viscosity at initial condition, cp;
$S_{or}$ = Residual oil saturation, fraction;
$p_i$ = Initial reservoir pressure, psi.

## PROGRAM DESCRIPTION

The program consists of only a main program and computes values for bulk reservoir volume, original oil in place, primary reserve, and recovery factor. It is emphasized that the reserve in this program is referred to as physically recoverable quantity by a known mechanism and not necessarily the economically recoverable quantity. The recovery factor is estimated for the reservoir as a whole, which may be unrealistic for a particular portion of the reservoir (lease) due to fluid migration (lease drainage).

Options are available for the selection of volumetric computation as well as recovery factor calculations, which take into consideration the type of recovery mechanism.

---

**Input Formats**

| Card | Format | Content |
|---|---|---|
| 1 | 20A4 | XN(I) |
| 2 | 3I1 | ITYPE, IREC, IVOL |
| 3<br>(Required if<br>  IVOL ≠ 2) | I2, 2F5.2, 3F5.3 | NC, T, H, APHI, ASWI, BOI |
| 4<br>(Required if<br>  IVOL ≠ 2) | 8F10.2 | A(I) |
| 5<br>(Required if<br>  IVOL = 2) | F10.2, F5.2, 3F5.3 | A(1), H, APHI, ASWI, BOI |
| 6<br>(Required if<br>  IREC = 2 and<br>  ITYPE = 1) | F5.4 | FGS |
| 7<br>(Required if<br>  IREC = 2 and<br>  ITYPE = 2) | F5.3, F5.0 | ROVIS, APERM |
| 8<br>(Required if<br>  IREC = 1 and<br>  ITYPE = 1) | 3F5.0, 2F5.2 | PB, PA, APERM, BOB,<br>VISOB |
| 9<br>(Required if<br>  IREC = 1 and<br>  ITYPE = 2) | 3F5.0, 3F5.2 | PI, PA, APERM, VISOI,<br>VISWI, BOI |

## Description of Input Parameters

XN(I) = Uses up to 80 alphanumeric characters for project identification;

ITYPE = Reservoir mechanism flag: 1 = depletion drive, 2 = water drive;

IREC = Type of recovery factor calculation to be performed: 1 = Arps correlation, 2 = residual oil correlation;

IVOL = Type of volumetric calculation desired: 1 = utilize Isopach map data, 2 = utilize uniformly thick circular reservoir data;

NC = Number of contour intervals;

T = Contour interval, ft;

H = Thickness of the volume enclosed between the last contour and the maximum thickness contour with zero area, ft.

APHI = Average reservoir porosity, fraction;

ASWI = Average initial water saturation, fraction;

BOI = Initial oil formation volume factor, bbl/STB;

A(I) = Area enclosed by successive contour, AC (must correspond with NC);

A(1) = Areal extent of the circular reservoir, AC;

H = Reservoir thickness (net), ft;

FGS = Reservoir gas saturation at abandonment, fraction; default value is 0.304;

ROVIS = Reservoir oil viscosity, cp;

APERM = Average reservoir permeability, md;

PB = Bubble point pressure, psi;

PA = Abandonment pressure, psi;

BOB = Oil formation volume factor at bubble point, bbl/STB;

VISOB = Reservoir oil viscosity at bubble point, cp;

PI = Initial reservoir pressure, psi;

VISOI = Reservoir oil viscosity at initial condition, cp;

VISWI = Viscosity of water at initial condition, cp.

## EXAMPLE PROBLEM

Determine original oil in place and recoverable reserve for a water drive reservoir with the following information. Use the Arps correlation for recovery factor.

| Contour | Area (Acre) | Contour Interval (ft) |
|---|---|---|
| 1 | 450 | 5 |
| 2 | 375 | 5 |
| 3 | 303 | 5 |
| 4 | 231 | 5 |
| 5 | 154 | 5 |
| 6 | 74 | 4 |

Porosity = 28.2%,
Water saturation = 35%,
Initial oil formation volume factor = 1.1 bbl/STB,
Initial reservoir pressure = 1986 psia,
Abandonment pressure = 800 psia,
Average reservoir permeability = 250 md,
Oil viscosity (reservoir condition) = 1.31 cp,
Water viscosity (aquifer) = 0.54 cp.

**INPUT LISTING**

FORTRAN CODING FORM

UU/CC  NAME  DATE

PROGRAM

ROUTINE

0 = ZERO   1 = ONE   Z = TWO
∅ = ALPHA O   I = ALPHA I   Z = ALPHA Z

| CARD | STATEMENT NUMBER 1 2 3 4 5 | 6 | 7 ... |
|------|---|---|---|

1. EXAMPLE PROBLEM  PROJECT 3  DATA SET 1

2. 214

3. 6.5.0014.00 12.82  350 4.1100

4. 45 0  375  30 3  231 1514

9. 198.6  8000  250 11 31 1514 1110  74

## SOURCE LISTING

```
      C     THIS PROGRAM CALCULATES ORIGINAL OIL IN PLACE BY VOLUMETRIC METHOD
      C     AND CALCULATES RECOVERABLE OIL BY PRIMARY DEPLETION BASED UPON
      C     TYPE OF RESERVOIR DRIVE AND CALCULATION DESIRED.
      C
            DIMENSION A(100),V(100),XN(20)
      C
      C     ITYPE=TYPE OF RESERVOIR DRIVE, 1=DEPLETION, 2=WATER DRIVE
      C     IREC=TYPE OF RECOVERY CALCULATION, 1=ARP'S STATISTICAL CORRELATION
      C     2=SIMPLE STATISTICAL OBSERVATION OF RESIDUAL OIL SATURATION RELATION
      C
      C     IVOL=TYPE OF VOLUMETRIC CALCULATION DESIRED, 1=ISOPACH MAP DATA,
      C     2=GIVEN DRAINAGE AREA AND NET THICKNESS DATA UTILIZED
 1000 CONTINUE
            READ(5,61)(XN(I),I=1,20)
   61 FORMAT(20A4)
            READ(5,1)ITYPE,IREC,IVOL
    1 FORMAT(3I1)
            IF(IVOL.EQ.2) GO TO 21
      C     NC=NUMBER OF CONTOUR INTERVALS
      C     T=CONTOUR INTERVAL, FT.
      C     A(I)=AREA UNDER EACH CONTOUR, STARTING WITH ZERO THICKNESS, ACRE
      C     H=THICKNESS OF THE VOLUME ENCLOSED BETWEEN THE LAST CONTOUR AND
      C     THE MAXIMUM THICKNESS CONTOUR WITH ZERO AREA
      C
      C     APHI=AVG. RESERVOIR POROSITY FRACTION
      C     ASWI=AVG. INITIAL WATER SATURATION FRACTION
      C     BOI=INITIAL OIL F.V.F , RESERVOIR BBL/STB
      C
            READ(5,2)NC,T,H,APHI,ASWI,BOI
    2 FORMAT(I2,2F5.2,3F5.3)
            READ(5,3)(A(I),I=1,NC)
    3 FORMAT(8F10.2)
            A(NC+1)=0.
            SUMV=0.
            DO 10 J=1,NC
            IF(J.EQ.NC)T=H
            V(J+1)=(T/3.)*(A(J)+A(J+1)+SQRT(A(J)*A(J+1)))
            SUMV=SUMV+V(J+1)
   10 CONTINUE
            OOIP=7758.*SUMV*APHI*(1.-ASWI)/BOI
            GO TO 12
   21 CONTINUE
            READ(5,4)A(1),H,APHI,ASWI,BOI
    4 FORMAT(F10.2,F5.2,3F5.3)
            SUMV=7758.*H*A(1)
            OOIP=7758.*H*A(1)*APHI*(1.-ASWI)/BOI
   12 CONTINUE
            IF(IREC.EQ.2.AND.ITYPE.EQ.1)GO TO 50
            IF(IREC.EQ.2.AND.ITYPE.EQ.2)GO TO 60
            IF(IREC.EQ.1.AND.ITYPE.EQ.1)GO TO 70
            IF(IREC.EQ.1.AND.ITYPE.EQ.2)GO TO 80
      C
      C     FGS=FINAL GAS SATURATION AT ABANDONMENT FOR DEPLETION TYPE RESERVE
      C     BOA=F.V.F AT ABANDONMENT
      C
   50 CONTINUE
            READ(5,18)FGS
   18 FORMAT(F5.4)
            IF(FGS.EQ.0.)FGS=.304
            RF=(1.-(BOI*(1.-ASWI-FGS)/(BOA*(1.-ASWI))))
            GO TO 100
      C
      C     ROVIS=RESERVOIR OIL VISCOSITY(INITIAL) CP
      C     ADERM=AVG*RESERVOIR PERMEABILITY,MD
      C
   60 READ(5,5)ROVIS,APERM
    5 FORMAT(F5.3,F5.0)
            IF(ROVIS.GE.0.2.AND.ROVIS.LT.0.5)SOR=30.+6.666*(ROVIS-0.2)
            IF(ROVIS.GE.0.5.AND.ROVIS.LT.1.0)SOR=32.+5.*(ROVIS-0.5)
            IF(ROVIS.GE.1.0.AND.ROVIS.LT.2.0)SOR=34.5+2.5*(ROVIS-1.)
            IF(ROVIS.GE.2.0.AND.ROVIS.LT.5.0)SOR=37.+1.1666*(ROVIS-2.)
            IF(ROVIS.GE.5.0.AND.ROVIS.LT.10.)SOR=40.5+0.6*(ROVIS-5.)
            IF(ROVIS.GE.10..AND.ROVIS.LT.20.)SOR=43.5+0.3*(ROVIS-10.)
            IF(ROVIS.GT.20.)WRITE(6,6)ROVIS
    6 FORMAT(//,3X,'RESERVOIR OIL VISCOSITY IS OUTSIDE THE RANGE',3X,
     *F6.2,' THE RANGE IS 20.')
            IF(APERM.LE.50.)DEV=12.
            IF(APERM.GE.50.0.AND.APERM.LT.100.)DEV=12.-0.06*(APERM-50.)
            IF(APERM.GE.100..AND.APERM.LT.200.)DEV=9.-0.03*(APERM-100.)
            IF(APERM.GE.200..AND.APERM.LT.500.)DEV=6.-.0133*(APERM-200.)
```

```
      IF(APERM.GE.500..AND.APERM.LT.1000.)DEV=2.-.006*(APERM-500.)
      IF(APERM.GE.1000..AND.APERM.LT.2000.)DEV=-1.-.0035*(APERM-1000.)
      IF(APERM.GE.2000..AND.APERM.LT.5000.)DEV=-4.5-.00133*(APERM-200
     *0.)
      IF(APERM.GT.5000.)DEV=-8.5
      CSOR=SOR+DEV
      RF=(1.-ASWI-CSOR)/(1.-ASWI)
      GO TO 100
C
C     PB=BUBBLE POINT PRESSURE, PSI
C     PA=ABONDONMENT PRESSURE, PSI
C     APERM=AVG. RESERVOIR PERMEABILITY, MD
C     BOB=R.V.F AT BUBBLE POINT PRESSURE
C     VISOB=RESERVOIR OIL VISCOSITY AT BUBBLE POINT
   70 CONTINUE
      READ(5,11)PB,PA,APERM,BOB,VISOB
   11 FORMAT(3F5.0,2F5.2)
      RF=0.41815*(APHI*(1.-ASWI)/BOB**.1611)*((APERM/VISCB)**.0979)
     **(ASWI**0.3722)*((PB/PA)**0.1741)




      GC TC 100
C
C     PI=INITIAL RESERVOIR PRESSURE
C     PA=ABANDOMENT PRESSURE
C     VISOI=INITIAL RESERVOIR OIL VISCOSITY
C     VISWI=WATER VISCCSITY UNDER INITIAL RESERVOIR CONDITION
C     BOI=INITIAL F.V.F
C     APERM=AVG. RESERVCIR PERMEABILITY
C
   80 CONTINUE
      READ(5,13)PI,PA,APERM,VISOI,VISWI,BOI
   13 FORMAT(3F5.0,3F5.2)
      RF=0.5489*((APHI*(1.-ASWI)/BOI)**0.0422*((VISWI*APERM/VISOI)
     ***(-0.077))*(ASWI**(-0.1903))*((PI/PA)**(-0.2159)))
  100 CONTINUE
      WRITE(6,71)(XN(I),I=1,20)
      RESERV=OCIP*RF
   71 FORMAT(20A4)
      WRITE(6,150)SUMV,APHI,ASWI,OCIP,RESERV,RF
  150 FORMAT(///,3X,'BULK VOLUME OF RESERVOIR ROCK ',F10.0,
     *' ACRE-FT',//,'    AVG. POROSITY ',F26.3, ' FRACTIONAL',//
     *'    AVG. INITIAL CIL SATURATION ',F12.3,' FRACTIONAL',//
     *,'    ORIGINAL-OIL-IN-PLACE ',F21.0,' STB',//,'    PRIMARY',
     *' RESERVE ',F27.0,' STB',//,'    RECOVERY FACTOR ',F23.2)
 2000 CONTINUE
      STOP
      END
//GO.SYSIN DD *
EXAMPLE PROBLEM     PROJECT 3     DATA SET 1
211
 6 5.00 4.00 .282 .3501.10
     450.        375.        303.        231.       154.      74.
1986. 800. 250. 1.31   .54 1.10
/*
```

## OUTPUT LISTING

EXAMPLE PROBLEM     PROJECT 3     DATA SET 1

| | |
|---|---|
| BULK VOLUME OF RESERVOIR ROCK | 6695. ACRE-FT |
| AVG. POROSITY | 0.282 FRACTIONAL |
| AVG. INITIAL OIL SATURATION | 0.350 FRACTIONAL |
| ORIGINAL-OIL-IN-PLACE | 8655041. STB |
| PRIMARY RESERVE | 3093302. STB |
| RECOVERY FACTOR | 0.36 |

**BIBLIOGRAPHY**

Arps, J. J.: "Reasons for Differences in Recovery Efficiency," paper SPE 2068, Petroleum Transaction Reprint Series No. 3, SPE, Dallas (1970) 49–54.

Craft, B. C., and Hawkins, M. F.: *Applied Petroleum Reservoir Engineering,* Chemical Engineering Series, Prentice-Hall, Englewood Cliffs, N.J. (1959).

Craze, R. C., Buckley, S. E.: "A Factual Analysis of the Effects of Well Spacing on Oil Recovery," *Drill. and Prod. Prac.,* API, **144** (1945); and Arps, J. J., *Trans.,* AIME (1956) **294,** 182.

# 5 ESTIMATION OF INITIAL OIL IN PLACE BY THE MATERIAL BALANCE METHOD FOR A SOLUTION GAS DRIVE RESERVOIR

**PURPOSE**

The initial oil in place for a solution gas/depletion drive reservoir is estimated, utilizing the generalized material balance equation and performance data. No distinction is made for calculations above or below bubble point conditions. However, for calculations below bubble point, the rock and water compressibility values are not important.

For a closed volumetric reservoir (no fluid migration in or out), the original oil-in-place value calculated for different times should agree within a few percent and tend to be constant. However, if there is fluid migration into the reservoir area—that is, water influx from an adjoining aquifer or gas influx from the gas cap—then the value of original oil in place with time should be increasing. None of these values computed assuming depletion drive is true under such conditions and should be disregarded.

**METHOD**

$$N = \frac{N_p[B_o + B_g(R_p - R_s)] + W_p}{B_o + (R_{si} - R_s)B_g + B_{oi}(C_f + C_w S_w)\dfrac{\Delta p}{1 - S_w}}, \tag{5.1}$$

where

$N$ = Original oil in place, STB;
$N_p$ = Oil produced (cumulative), STB;
$B_o$ = Oil formation volume factor, res. bbl/STB;
$B_g$ = Gas formation volume factor, bbl/SCF;
$R_p$ = Produced gas-oil ratio, SCF/STB;

41

$R_s$ = Solution gas-oil ratio, SCF/STB;

$W_p$ = Cumulative water produced, bbl;

$R_{si}$ = Solution gas-oil ratio at initial condition, SCF/STB;

$B_{oi}$ = Oil formation volume factor at initial condition, bbl/STB;

$C_f$ = Reservoir rock compressibility, vol/vol/psi;

$C_w$ = Water compressibility, vol/vol/psi;

$S_w$ = Water saturation, fraction;

$\Delta P$ = Pressure decline, psi.


## PROGRAM DESCRIPTION

The program consists of a simple main program and performs one calculation at each pressure step using equation (5.1). Up to 20-pressure-step performance data may be utilized.

The development of the material balance equation is based on the assumption that the reservoir is continuous and uniform and that the fluid withdrawals are uniformly distributed throughout. Although few reservoirs are uniform, reasonable results are obtained using average conditions for the pool.


### Input Formats

| Card | Format | Content |
|------|--------|---------|
| 1 | 20A4 | A(I) |
| 2 | 20A4 | B(I) |
| 3 | 2F5.0, F5.3, F5.0, F5.3 | PRI, BPP, BOI, RSI, SWI |
| 4 | 3F8.6, I2 | CF, CO, CW, N |
| 5–N | F5.0, F10.0, F5.0, F10.0, F5.3, F5.0, F6.5 | C2(I), C3(I), C4(I), C5(I), C6(I), C7(I), C9(I) |


### Description of Input Parameters

A(I) = Uses up to 80 alphanumeric characters for project identification;

B(I) = Uses up to 80 alphanumeric characters for project identification;

PRI = Initial reservoir pressure, psi;

BPP = Bubble point pressure, psi;

BOI = Oil formation volume factor at initial condition, bbl/STB;

RSI = Solution oil-gas ratio at initial condition, SCF/STB;

SWI = Initial water saturation, fraction;

CF = Rock compressibility, vol/vol/psi;

CO = Oil compressibility, vol/vol/psi;

Table 5.1 Pressure and Production Data for a Solution-Drive Type Reservoir

| Pressure (psia) | Cumulative Production (MSTB) | Cumulative Gas-Oil Ratio (SCF/STB) | Solution Gas-Oil Ratio (SCF/bbl) | Oil Formation Volume Factor (bbl/STB) | Gas Formation Volume Factor (SCF/STB) |
|---|---|---|---|---|---|
| 1800 | 0 | 0 | 577 | 1.268 | 0.00097 |
| 1482 | 2223 | 634 | 491 | 1.233 | 0.00119 |
| 1367 | 2981 | 707 | 460 | 1.220 | 0.00130 |
| 1053 | 5787 | 1034 | 375 | 1.186 | 0.00175 |

CW = Water compressibility, vol/vol/psi;
N = Number of time steps, integer value;
C2(I) = Weighted average reservoir pressure, psi;
C3(I) = Cumulative oil production, STB;
C4(I) = Cumulative produced gas-oil ratio, SCF/STB;
C5(I) = Cumulative water produced, bbl;
C6(I) = Oil formation volume factor, bbl/STB;
C7(I) = Solution gas-oil ratio, SCF/STB;
C9(I) = Gas formation volume factor, bbl/SCF.

**EXAMPLE PROBLEM**

Pressure and production data for a solution-drive-type reservoir are shown in table 5.1. Other information is as follows:

Initial reservoir pressure = 1800 psia,
Initial oil formation volume factor = 1.268 bbl/STB,
Initial solution gas-oil ratio = 577 SCF/STB,
Initial water saturation = 20%.

Determine original oil in place using material balance method.

**INPUT LISTING**

# FORTRAN CODING FORM

UU/CC   NAME
        DATE

PROGRAM
ROUTINE

0 = ZERO
0 = OH

1 = BL
1 = ALPHA 1

7 = 143
7 = ALPHA Z

| CARD | STATEMENT NUMBER | | |
|---|---|---|---|
| 1. | T E S T | E X A M P L E   P R O B L E M   4 | M O R I G A N L T . O W N        W E S T   V I R G I N I A |
| 2. | T E S T E D | | |
| 3. | 1 8 0 0 . 1 | 8 0 0 . 1 . . 2 6 8 . 5 7 7 . . . 2 0 0 | |
| 4. | 1 . 0 L | 0 . . . . . 1 . . 0 L        4 . | |
| 5.1 | 1 8 0 0 . | 0 0 0 8 0 0 0 . . . 0 . | 0 . . 1 . . 2 6 8 . 5 7 7 . . 0 0 0 9 7 |
| 5.2 | 1 4 8 2 . | 2 2 3 0 0 0 . . 6 3 4 . | 0 . . 1 . . 2 3 3 . 4 9 . . . 0 0 1 1 9 |
| 5.3 | 1 3 6 7 . | 2 9 8 1 0 0 0 . . 7 0 7 . | 0 . . 1 . . 2 2 0 . 4 6 0 . . 0 0 1 3 0 |
| 5.4 | 1 0 5 3 . | 5 7 8 7 0 0 0 . . 0 3 4 . | 0 . . 1 . . 1 8 6 . 3 7 5 . . . 0 0 1 7 5 |

## SOURCE LISTING

```
C        THIS PROGRAM CALCULATES THE ORIGINAL OIL IN PLACE UTILIZING
C        GENERALIZED MATERIAL BALANCE EQUATION FOR SOLUTION DRIVE
C        RESERVOIRS WITHOUT GAS-CAP OR WATER DRIVE.
C
C        C2(I)=RESERVOIR PRESSURE, PSI
C        C3(I)=CUM.OIL PRODUCTION, BBL
C        C4(I)=CUM.GOR GAS, SCF
C        C5(I)=CUM.WATER PRODUCTION, BBL
C        C6(I)=OIL FVF, RES.BBL/STB
C        C7(I)=SOLN.GAS-OIL RATIO, SCF/STB
C        C8(I)=ORIGINAL OIL-IN-PLACE, STB
C        C9(I)=GAS FVF, RES.BBL/STB
C        N=NUMBER OF DATA POINT INCLUDING THE INITIAL CONDITION
C
         DIMENSION C1(20),C2(20),C3(20),C4(20),C5(20),C6(20),C7(20),C8(20),
        *C9(20),C10(20),A(20),B(20)
         WRITE(6,1)
   1     FORMAT(//,3X,'DETERMINATION OF ORIGINAL OIL IN PLACE')
         WRITE(6,2)
   2     FORMAT(/,3X,'BY GENERALIZED MATERIAL BALANCE APPROACH')
         WRITE(6,3)
   3     FORMAT(//,3X,'FOR SOLUTION GAS RESERVOIR WITH NO GAS-CAP OR WATER'
        *,'DRIVE')
C
C        CF=ROCK COMPRESSIBILITY, VOL/VOL/PSI
C        CO=OIL COMPRESSIBILITY, VOL/VOL/PSI
C        CW=WATER COMPRESSIBILITY, VCL/VOL/PSI
C        SW=INITIAL WATER SATURATION, FRACTION OF PORE SPACE
C
         READ(5,4)(A(I),I=1,20)
         READ(5,4)(B(I),I=1,20)
   4     FORMAT(20A4)
         READ(5,5)PRI,BPP,BOI,RSI,SWI
   5     FORMAT(2F5.0,F5.3,F5.0,F5.3)
         READ(5,7)CF,CO,CW,N
   7     FORMAT(3F8.6,I2)
         DO 10 I=1,N
         READ(5,8)C2(I),C3(I),C4(I),C5(I),C6(I),C7(I),C9(I)
   8     FORMAT(F5.0,F10.0,F5.0,F10.0,F5.3,F5.0,F6.5)
  10     CONTINUE
         WRITE(6,12)(A(I),I=1,20)
  12     FORMAT(//,3X,20A4)
         WRITE(6,12)(B(I),I=1,20)
         WRITE(6,11)
  11     FORMAT(//,3X,'INPUT VARIABLE LISTING')
         WRITE(6,13)PRI
         WRITE(6,14)BOI
         WRITE(6,15)RSI
         WRITE(6,16)SWI
         WRITE(6,17)CF
         WRITE(6,18)CO
         WRITE(6,19)CW
  13     FORMAT(/,3X,'INITIAL RESERVOIR PRESSURE',3X,F6.0,' PSIA')
  14     FORMAT(  3X,'INITIAL OIL FVF          ',3X,F6.3,' RB/STB')
  15     FORMAT(  3X,'INITIAL SOLN. GOR        ',3X,F6.0,' SCF/STB')
  16     FORMAT(  3X,'INITIAL WATER SATURATION ',3X,F6.3,' FRACTION')
  17     FORMAT(  3X,'ROCK COMP.FACTOR         ',3X,F6.5,' VOL/VOL/PSI')
  18     FORMAT(  3X,'OIL  COMP.FACTOR         ',3X,F6.5,' VOL/VOL/PSI')
  19     FORMAT(  3X,'WATER COMP.FACTOR        ',3X,F6.5,' VCL/VOL/PSI')
         WRITE(6,20)
  20     FORMAT(/,3X,'PRESS.',3X,'CUM.OIL',3X,'CUM.GOR',3X,'CUM.WATER',
        *3X,'OIL FVF',3X,'SOLN.GOR',3X,'GAS FVF')
         DO 25 I=1,N
         WRITE(6,21)C2(I),C3(I),C4(I),C5(I),C6(I),C7(I),C9(I)
  21     FORMAT(F9.0,F11.0,F10.0,F12.0,F10.3,F11.0,F10.6)
  25     CONTINUE
         DO 30 I=2,N
         IF(C2(I).GE.BPP)CO=(C6(I)/BOI-1.)/C10(I)
         C8(I)=(C3(I)*(C6(I)+C9(I)*(C4(I)-C7(I)))+C5(I))/((C6(I)
        *-C6(1))+(C9(I)*(C7(1)-C7(I)))+(C6(1)*(SWI*CW+CF)*C10(I)/(1.-SWI)))
  30     CONTINUE
         WRITE(6,31)
  31     FORMAT(//,3X,'PRESSURE',3X,'CUM.OIL PROD.',3X,'CUM.GOR',3X,
        *'ORIGINAL OIL IN PLACE')
         WRITE(6,32)
  32     FORMAT(5X,'PSI',12X,'BBL',7X,'SCF/BBL',12X,'STB')
         DO 40 I=2,N
         WRITE(6,41)C2(I),C3(I),C4(I),C8(I)
  41     FORMAT(F11.0,F16.0,F10.0,F19.0)
  40     CONTINUE
```

```
     STOP
     END
//GO.SYSIN DD *
TEST EXAMPLE PROBLEM 4        MORGANTOWN           WV    P 64
TESTED
1800.1800.1.268 577.  .200
 .0      .0      .0             4
1800.0.       0.    0.          1.268577.  .00097
1482.2223000.    634. 0.        1.233491.  .00119
1367.2981000.    707. 0.        1.220460.  .00130
1053.5787000.    1034.0.        1.186375.  .00175
/*
//
```

## OUTPUT LISTING

```
DETERMINATION OF ORIGINAL OIL IN PLACE

BY GENERALIZED MATERIAL BALANCE APPROACH

FCR SOLUTION GAS RESERVOIR WITH NO GAS-CAP OR WATERDRIVE

TEST EXAMPLE PROBLEM 4        MCRGANTOWN           WV    P 64

TESTED

INPUT VARIABLE LISTING
INITIAL RESERVOIR PRESSURE     1800. PSIA
INITIAL OIL FVF                1.268 RB/STB
INITIAL SOLN. GOR              577. SCF/STB
INITIAL WATER SATURATION       0.200 FRACTION
ROCK COMP.FACTOR               .0    VOL/VOL/PSI
OIL  COMP.FACTOR               .0    VOL/VOL/PSI
WATER COMP.FACTOR              .0    VOL/VOL/PSI
```

| PRESS. | CUM.OIL | CUM.GOR | CUM.WATER | OIL FVF | SOLN.GOR | GAS FVF |
|---|---|---|---|---|---|---|
| 1800. | 0. | 0. | 0. | 1.268 | 577. | 0.000970 |
| 1482. | 2223000. | 634. | 0. | 1.233 | 491. | 0.001190 |
| 1367. | 2981000. | 707. | 0. | 1.220 | 460. | 0.001300 |
| 1053. | 5787000. | 1034. | 0. | 1.186 | 375. | 0.001750 |

| PRESSURE PSI | CUM.OIL PROD. BBL | CUM.GOR SCF/BBL | ORIGINAL OIL IN PLACE STB |
|---|---|---|---|
| 1482. | 2223000. | 634. | 46320752. |
| 1367. | 2981000. | 707. | 44130544. |
| 1053. | 5787000. | 1034. | 49860864. |

## BIBLIOGRAPHY

Craft, B. C., and Hawkins, M. F.: *Applied Petroleum Reservoir Engineering,* Prentice-Hall, Englewood Cliffs, N.J. (1959).

Slider, H. C.: *Practical Petroleum Reservoir Engineering Methods,* The Petroleum Publishing Company, Tulsa (1976).

# 6 DETERMINATION OF ORIGINAL OIL IN PLACE BY THE MATERIAL BALANCE METHOD FOR A RESERVOIR WITH INITIAL GAS CAP AND NO WATER INFLUX

## PURPOSE

The program performs a set of material balance calculations at different time steps for a reservoir with an initial gas cap. In the formulation of the material balance equation, it is assumed that the reservoir produces under a combined (solution gas and gas cap) expansion drive. Gravity effects and water influx are assumed negligible. The water and rock compressibilities are also neglected. For the purpose of material balance calculations, the gas-cap gas volume as well as the oil-band oil volume must be known. These volumes usually are computed using the volumetric method. The original oil-in-place determination by the material balance procedure is often considered a secondary check.

If the ratio of the gas-cap gas volume to the oil-band oil volume is around 0.5 or higher, the energy contributed by the gas-cap expansion is significant compared to the solution gas expansion energy. Therefore, a slight error in the average reservoir pressure (10 to 20 psi) may cause significant error in the computed values of oil in place, due to over- or underexpansion of the gas cap. The exclusion of rock and water compressibilities, however, does not materially affect the computed values.

In the formulation of the material balance equation, the water influx term is neglected, assuming no water influx. However, if water influx is not negligible in reality, then the computed values of initial oil in place by this program will be high.

## METHOD

$$N = \frac{N_p[B_o + B_g(R_p - R_s)] + W_p}{B_o + B_g(R_{si} - R_s) + mB_{oi}\left(\dfrac{B_g}{B_{gi}} - 1\right) - B_{oi}} \tag{6.1}$$

47

$$R_p = \frac{G_p}{N_p} = \frac{\sum \Delta N_p R_{Iavg}}{N_p} = \frac{\sum\limits_{i=1}^{i=j} (\Delta N_p) [R_{i(g)} + R_{I(g-1)i}]0.5}{N_p} \tag{6.2}$$

$$W_p = \sum \Delta W_p = \sum\limits_{i=1}^{i=j} \left[ \frac{f_{wavg} \times \Delta N_p}{1 - f_{wavg}} \right]_i \tag{6.3}$$

where

$N$ = Initial oil in place, STB;
$N_p$ = Cumulative oil production, STB;
$B_o$ = Oil formation volume factor, res. bbl/STB;
$B_g$ = Gas formation volume factor, res. bbl/SCF;
$R_p$ = Cumulative produced gas-oil ratio, SCF/STB;
$R_s$ = Solution gas-oil ratio, SCF/STB;
$W_p$ = Cumulative water production, bbl;
$R_{si}$ = Initial solution gas-oil ratio, SCF/STB;
$m = \dfrac{\text{Initial pore space occupied by free gas in the gas cap}}{\text{Initial pore space occupied by oil in the oil band}}$, ratio;
$B_{oi}$ = Initial oil formation volume factor, bbl/STB;
$B_{gi}$ = Initial gas formation volume factor, bbl/SCF;
$G_p$ = Cumulative gas produced, SCF;
$\Delta N_p$ = Oil produced in a pressure interval, STB;
$\Delta W_p$ = Water produced in a pressure interval, bbl;
$R_{Iavg}$ = Average instantaneous gas-oil ratio, SCF/STB;
$f_{wavg}$ = Average water cut, fraction.

**PROGRAM DESCRIPTION**
The program consists of a simple FORTRAN main program and performs one calculation at each pressure step using equations (6.1) through (6.3). The accuracy of the answers obtained, of course, depends on the reliability and sufficiency of the data. However, even under favorable conditions, an error in the range of 15 to 20% is possible. Up to 20-pressure-step performance data may be utilized. The original oil-in-place values computed at each pressure step should be within a few percent and should tend to be constant.

**Input Format**

| Card | Format | Content |
|------|--------|---------|
| 1 | 20A4 | A(I) |
| 2 | F5.0, F5.3, I2, F5.3, F5.0, F8.7 | PRI, XMI, N, BOI, RSI, BGI |
| 3–N | F5.4, F6.0, F13.4, F8.3, F10.7, F5.1 | C1(I), C2(I), C3(I), C4(I), C6(I), C5(I) |

### Description of Input Parameters

A(I) = Uses up to 80 alphanumeric characters for project identification;
PRI = Initial reservoir pressure, psi;
XMI = Ratio of initial gas-cap gas pore volume to oil pore volume;
N = Number of pressure steps to be evaluated;
BOI = Initial oil formation volume factor, bbl/STB;
RSI = Initial solution gas-oil ratio, SCF/STB;
BGI = Initial gas formation volume factor, bbl/SCF;
C1(I) = Average reservoir pressure at pressure step I, psi;
C2(I) = Producing gas-oil ratio at pressure step I, SCF/STB;
C3(I) = Cumulative oil produced at pressure step I, STB;
C4(I) = Two-phase formation volume factor at pressure step I, bbl/STB;
C6(I) = Gas formation volume factor at pressure step I, bbl/SCF;
C5(I) = Water cut at pressure step I, percent.

### EXAMPLE PROBLEM

Pressure, production, and fluid data for a reservoir are given in table 6.1. Other data on this reservoir are as follows:

Initial reservoir pressure = 2920 psi;
Initial oil in place by volumetric method = 223 million STB;
Initial reservoir pore volume occupied by free gas = 46.4 million bbl;
Reservoir temperature = 211°F;
Initial oil formation volume factor = 1.454 bbl/STB.

Determine the original oil in place by the material balance method.

**Table 6.1 Pressure, Production, and Fluid Property Data**

| Pressure | Producing Gas-Oil Ratio | Cumulative Oil Production (MMSTB) | Two-Phase Formation Volume Factor | Water Cut (percent) | Gas Formation Volume Factor |
|---|---|---|---|---|---|
| 2920 | 780 | 0 | 1.454 | 0 | 0.954 |
| 2740 | 1150 | 4.1 | 1.477 | 0.6 | 1.004 |
| 2560 | 1885 | 8.3 | 1.506 | 4.7 | 1.072 |
| 2300 | 2670 | 12.7 | 1.565 | 7.3 | 1.194 |
| 2050 | 3713 | 17.1 | 1.648 | 8 | 1.347 |
| 1800 | 4480 | 21.7 | 1.757 | 6.7 | 1.550 |
| 1500 | 4320 | 26.3 | 1.956 | 7.2 | 1.905 |
| 1220 | 4020 | 31.2 | 2.276 | 11 | 2.390 |

**INPUT LISTING**

FORTRAN CODING F...

UU/CC   NAME
        DATE

PROGRAM
ROUTINE

```
                      D = ZERO          1 = ONE
                      Ø = ALPHA O       I = ALPHA I
```

| CARD | STATEMENT NUMBER | | |
|---|---|---|---|
| 1. | | EXAMPLE PROBLEM 15          SALT LAKE CITY          UTAH |
| 2. | 2,9,2,0. | .1,4,3 .8,1,.1,4,5,4 .7,8 01..1010954,0 |
| 3-1 | 2,9,2,0. | 7,8,0. .01 .1,4,5,4 .10109594 0..10 |
| 3-2 | 2,7,4,0. | 1,1,5,0. 4,.0 .1,4,7,7 .1011010 4 0..16 |
| 3-3 | 2,5,6,0. | 1,8,8,5. 8,.3 .1,5,0,6 .100 1017 2 4..17 |
| 3-4 | 2,3,0,0. | 2,6,7,0. 1,2,.7 .1,5,6,5 .100 1194 7..13 |
| 3-5 | 2,0,5,0. | 3,7,1,3. 1,7,.1 .1,6,4,8 .1001347 8..10 |
| 3-6 | 1,8,0,0. | 4,4,8,0. 2,1,.7 .1,7,5,7 .100155,0 6..17 |
| 3-7 | 1,5,0,0. | 4,3,2,0. 2,6,.3 .1,9,5,6 .1001905 7..12 |
| 3-8 | 1,2,0,0. | 4,0,2,0. 3,1,.2 .2,2,7,6 .1001239 6 11..10 |

**SOURCE LISTING**

```
C          THIS PROGRAM DETERMINES THE ORIGINAL OIL-IN-PLACE FOR RESERVOIR
C          WITH INITIAL GAS-CAP (NO WATER DRIVE) BY MATERIAL BALANCE APPROACH
C
C          PRI=INITIAL RESERVOIR PRESSURE, PSIA
C          XMI=RATIO OF INITIAL PORE-SPACE OCCUPIED BY FREE GAS TO THE PORE
C          SPACE OCCUPIED BY OIL FRACTION
C          N=NUMBER OF PRESSURE PRODUCTION DATA POINT AVAILABLE FOR
C          CALCULATION PURPOSES (SHOULD NOT EXCEED 20)
C          C1=RESERVOIR PRESSURE (STATIC STABILIZED), PSIA
C          C2=PRODUCING GOR, SCF/STB
C          C3=CUM.OIL PRODUCTION, MM STB
C          C27=OIL FORMATION VOLUME FRACTION, RESERVOIR BBL/STB
C          C28=SOLUTION GAS-OIL RATIO, SCF/STB
C          C5=WATER CUT PERCENT
C          C6=GAS FORMATION VOLUME FACTOR, RESOIR BBL/SCF



C          A=ALPHA FIELD FOR PROBLEM IDENTIFICATICN CR HEADING
C          C4=TWO PHASE FVF, RES.BBL/STB
C
C
           DIMENSION C1(20),C2(20),C3(20),C4(20),C5(20),C6(20),C7(20),
          *C8(20),C9(20),C10(20),C11(20),C12(20),C13(20),C14(20),
          *C15(20),C16(20),C17(20),C18(20),C19(20),C20(20),C21(20),
          *C22(20),C23(20),C24(20),C25(20),C26(20),C27(20),C28(20),
          *A(20)
           READ(5,1)(A(I),I=1,20)
    1      FORMAT(20A4)
           READ(5,2)PRI,XMI,N,BOI,RSI,BGI
    2      FORMAT(F5.0,F5.3,I2,F5.3,F5.0,F8.7)
           DO 10 I=1,N
           READ(5,3)C1(I),C2(I),C3(I),C4(I),C6(I),C5(I)
    3      FORMAT(F5.0,F6.0,F13.4,F8.3,F10.7,F5.1)
   10      CONTINUE
           WRITE(6,25)
           WRITE(6,20)(A(I),I=1,20)
   25      FORMAT(/,3X,'DETERMINATION OF ORIGINAL-OIL-IN-PLACE FOR',
          *' RESERVOIR ',/,3X,'WITH INITIAL GAS-CAP (NO WATER INFLUX)',
          *' BY MATERIAL BALANCE EQUATION')
   20      FORMAT(//,3X,20A4)
           WRITE(6,50)
   50      FORMAT(///,3X,'INPUT VARIABLES LIST')
           WRITE(6,51)
   51      FORMAT(/,3X,'PRES.',3X,'PRD.GOR',3X,'CUM.CIL PROD.',3X,
          *'T.P-FVF',3X,'GAS FVF',3X,'WATER CUT')
           WRITE(6,52)
   52      FORMAT(3X,'PSIA',3X,'SCF/STB',4X,'MILLION   STB.',3X,'RBL/STB',
          *3X,'RBL/SCF',4X,'PERCENT')
           DO 53 I=1,N
           WRITE(6,54)C1(I),C2(I),C3(I),C4(I),C6(I),C5(I)
   54      FORMAT(3X,F5.0,3X,F7.0,3X,F8.0,8X,F7.3,3X,F7.5,3X,F9.2)
   53      CONTINUE
           WRITE(6,55)PRI,XMI
   55      FORMAT(/,3X,'INITIAL RESERVOIR PRESSURE',3X,F5.0,1X,'PSIA',
          */,3X,'RATIO OF GAS-CAP TO OIL PORE VOLUME',3X,F5.3)
           WRITE(6,56)
   56      FORMAT(///,3X,'CALCULATED VALUES OR OUTPUT LIST')
           SUM=0.
           C10(1)=0.
           C4(1)=C27(1)
           C15(1)=0.
           DO 100 I=2,N
           C7(I)=C3(I)-C3(I-1)
           C8(I)=0.5*(C2(I)+C2(I-1))
           C9(I)=C7(I)*C8(I)
           C10(I)=C10(I-1)+C9(I)
           C11(I)=C10(I)/C3(I)
           C12(I)=0.005*(C5(I)+C5(I-1))
           C13(I)=C7(I)*C12(I)
           C14(I)=C13(I)/(1.-C12(I))
           C15(I)=C15(I-1)+C14(I)
           C16(I)=C11(I)-RSI
           C17(I)=C6(I)*C16(I)
           C18(I)=C4(I)+C17(I)
           C19(I)=C18(I)*C3(I)
```

```
          C20(I)=C19(I)+C15(I)
          C21(I)=C4(I)-BOI
          C22(I)=C6(I)/BGI
          C23(I)=C22(I)-1.
          C24(I)=XMI*BOI*C23(I)
          C25(I)=C21(I)+C24(I)
          C26(I)=C20(I)/C25(I)
          SUM=SUM+C26(I)
    100 CONTINUE
          AVGCIP=SUM/FLOAT(N-1)
          WRITE(6,35)
     35 FCRMAT(/,3X,'PRES.',3X,'CUM.CIL',3X,'CUM.GAS',3X,'CUM.WATER',
        *3X,'ORIGINAL-OIL-IN-PLACE',/,3X,'PSIA',4X,'MM BBL',4X,'MM SCF',
        *6X,'MM BBL',13X,'MM STB')
          DC 200 I=2,N
          WRITE(6,40)C1(I),C3(I),C10(I),C15(I),C26(I)
     40 FORMAT(3X,F5.0,3X,F7.0,3X,F7.0,3X,F9.0,10X,F9.0)



    200 CCNTINUE
          WRITE(6,41)AVGCIP
     41 FORMAT(/,3X,'AVERAGE VALUE CF ORIGINAL-OIL-IN-PLACE',3X,F10.0,
        *' MM STB')
          STOP
          END
//GO.SYSIN DD *
EXAMPLE PROBLEM  5      SALT LAKE CITY, UTAH
2920.0.143081.454078C..0009540
2920.780.          0.          1.454   .000954   0.
2740.1150.         4.1         1.477   .001004   .6
2560.1885.         8.3         1.506   .001072   4.7
2300.2670.         12.7        1.565   .001194   7.3
2050.3713.         17.1        1.648   .001347   8.0
1800.4480.         21.7        1.757   .001550   6.7
1500.4320.         26.3        1.956   .001905   7.2
1200.4020.         31.2        2.276   .002390   11.0
/*
//
```

## OUTPUT LISTING

```
DETERMINATION OF ORIGINAL-CIL-IN-PLACE FOR RESERVOIR
WITH INITIAL GAS-CAP (NO WATER INFLUX) BY MATERIAL BALANCE EQUATION

EXAMPLE PROBLEM  5      SALT LAKE CITY, UTAH


INPUT VARIABLES LIST
```

| PRES. | PRD.GOR | CUM.OIL PROD. | T.P-FVF | GAS FVF | WATER CUT |
|---|---|---|---|---|---|
| PSIA | SCF/STB | MILLION STB. | RBL/STB | RBL/SCF | PERCENT |
| 2920. | 780. | 0. | 1.454 | 0.00095 | 0.0 |
| 2740. | 1150. | 4. | 1.477 | 0.00100 | 0.60 |
| 2560. | 1885. | 8. | 1.506 | 0.00107 | 4.70 |
| 2300. | 2670. | 13. | 1.565 | 0.00119 | 7.30 |
| 2050. | 3713. | 17. | 1.648 | 0.00135 | 8.00 |
| 1800. | 4480. | 22. | 1.757 | 0.00155 | 6.70 |
| 1500. | 4320. | 26. | 1.956 | 0.00191 | 7.20 |
| 1200. | 4020. | 31. | 2.276 | 0.00239 | 11.00 |

```
INITIAL RESERVOIR PRESSURE    2920. PSIA
RATIO OF GAS-CAP TO OIL PORE VOLUME    0.143
```

```
CALCULATED VALUES OR CUTPUT LIST
```

| PRES. PSIA | CUM.OIL MM BBL | CUM.GAS MM SCF | CUM.WATER MM BBL | ORIGINAL-OIL-IN-PLACE MM STB |
|---|---|---|---|---|
| 2740. | 4. | 3957. | 0. | 201. |
| 2560. | 8. | 10330. | 0. | 216. |
| 2300. | 13. | 20351. | 0. | 201. |
| 2050. | 17. | 34394. | 1. | 205. |
| 1800. | 22. | 53237. | 1. | 221. |
| 1500. | 26. | 73477. | 1. | 217. |
| 1200. | 31. | 93910. | 2. | 211. |

```
AVERAGE VALUE OF ORIGINAL-OIL-IN-PLACE          210. MM STB
```

# BIBLIOGRAPHY

Guerrero, E. T.: *Practical Reservoir Engineering,* The Petroleum Publishing Company, Tulsa (1968).

# 7    DETERMINATION OF OIL IN PLACE BY THE MATERIAL BALANCE METHOD FOR RESERVOIRS WITH PARTIAL WATER DRIVE (NO GAS CAP)

**PURPOSE**

The program computes a set of values for initial oil in place for a partial water drive reservoir with no gas cap. One calculation is performed for each time step and corresponding reservoir pressure and for cumulative production during the available performance history. Water influx computations are performed utilizing the Van Everdingen-Hurst unsteady-state equation. A set of response functions ($Q_T$) and corresponding dimensionless time ($T_D$) values are provided by the user. The selection of the $Q_T$ function set is based on the assumed size of the aquifer compared to the size of the reservoir.

Although the material balance calculations can be performed above the bubble point pressure, this program is designed to perform all computations below the bubble point pressure. At pressures above the bubble point and slightly below, the material balance calculations are highly sensitive to slight errors in pressures and fluid property data. High sensitivity to such errors makes it very difficult to obtain a reliable estimate above the bubble point pressure.

Computational results are also quite sensitive to the values of formation volume factors. Therefore, a very high degree of accuracy (6 or 7 decimal places) is suggested for these parameters. It is realized that such accuracy is not obtainable from the laboratory measurement. However, this can be achieved by expressing the formation volume factor and pressure relationship with an equation, which may then be solved to obtain the desired accuracy.

**METHOD**

$$N = \frac{N_p[B_t + B_g(R_p - R_{si})] + W_p - W_e}{B_t - B_{ti}}. \tag{7.1}$$

**55**

Rearranging equation (7.1), we may write

$$\frac{N_p[B_t + B_g(R_p - R_{si})] + W_p}{B_t - B_{ti}} = \frac{W_e}{B_t - B_{ti}} + N. \tag{7.2}$$

$$W_e = B\sum_{i=1}^{n} \Delta P_i Q_T(n + 1 - i). \tag{7.3}$$

Substitution of equation (7.3) in (7.2) yields

$$\frac{N_p[B_t + B_g(R_p - R_{si})] + W_p}{B_t - B_{ti}} = \frac{B\sum_{i=1}^{n} \Delta P_i Q_T(n + 1 - i)}{B_t - B_{ti}} + N. \tag{7.4}$$

Let

$$\frac{N_p[B_t + B_g(R_p - R_{si}) + W_p}{B_t - B_{ti}} = Y \tag{7.5}$$

and

$$\frac{\sum_{i=1}^{n} \Delta P_i Q_{T(n + 1 - i)}}{B_t - B_{ti}} = X. \tag{7.6}$$

Then substitution of equations (7.5) and (7.6) in (7.4) yields

$$Y = BX + N, \tag{7.7}$$

$$N = \Sigma x^2 \Sigma y - \Sigma \times \Sigma \times y/[(n\Sigma x^2 - (\Sigma x)^2)], \tag{7.8}$$

and

$$B = \frac{\Sigma y - nN}{\Sigma x}. \tag{7.9}$$

Equation (7.7) defines a straight line with a slope $B$ and intercept value $N$. $N$ is the original oil in place, and $B$ is the proportionality factor of the Van Everdingen-Hurst equation. The reduction of the material balance equation (7.1) to the form of equation (7.7) is important. This reduction allows one to verify the input data and, specifically, the assumption regarding the aquifer size and selection of the corresponding $Q_T$ function value set.

A reasonably constant computed value of $B$, particularly during the later stages, will indicate that reliable estimates and assumptions have been made. Otherwise, the assumptions should be changed and the computation repeated.

For equation (7.1) to (7.9),

$N$ = Initial oil in place, *STB;*

$N_p$ = Cumulative oil production, STB;

$B_t$ = Two-phase formation volume factor, bbl/STB;

$B_g$ = Gas formation volume factor, bbl/SCF;

$R_p$ = Cumulative produced gas-oil ratio, SCF/STB;

$R_{si}$ = Initial solution gas ratio, SCF/STB;

$B_{ti}$ = Initial two-phase formation volume factor, bbl/STB;

$W_p$ = Cumulative water production, bbl;

$W_e$ = Cumulative water influx, bbl;

$B$ = Water influx or proportionality constant, bbl/psi;

$\Delta P_i = P_i - P$ = pressure difference—that is, pressure drop occurring during successive time intervals from the initial pressure, psi;

$Q_T$ = Dimensionless water influx response function;

$n$ = Number of time increments.

## PROGRAM DESCRIPTION

The program consists of a FORTRAN main program and one subprogram. It performs one calculation at each time and pressure step using equations (7.5), (7.6), (7.8), and (7.9). At least 4-time-step performance data below the bubble point pressure should be available. Up to 20-time-step data can be provided.

The success of the procedure depends upon reliable fluid, rock, and production data as well as assumptions involved in the material balance equation and unsteady-state water influx term.

A set of discrete values of $Q_T$ versus $T_D$ is provided by the user. This program uses a SPLINE curve-fitting program to determine $Q_T$ values on a continuous basis for use in the computation of water influx values.

## Input Format

| Card | Format | Content |
|---|---|---|
| 1 | 20A4 | A(I) |
| 2 | 20A4 | B(I) |
| 3 | I2 | N |
| 4 | 3E10.4, F10.7 | AVPOR, APERMW, AVISW, CW |
| 5 | F10.0, 2F5.0, F7.5, F8.0, F3.0 | XNPB, PB, RSI, BOB, RW, RERW |
| 6–N | 3F6.0, 2F10.0, F7.0, 2F7.0, F9.7, F9.7 | C1(I), C2(I), C10(I), C3(I), C4(I), C6(I), C8(I), C9(I), C5(I), C7(I) |
| 7 | I3 | NN |
| 8–NN | 2E10.5 | TD(I), QT(I) |

### Description of Input Parameters

$A(I)$ = Uses up to 80 alphanumeric characters for project identification;
$B(I)$ = Uses up to 80 alphanumeric characters for project identification;
$N$ = Number of time-step data provided;
$AVPOR$ = Average reservoir porosity, fraction;
$APERMW$ = Average aquifer permeability, md;
$AVISW$ = Average water viscosity under reservoir conditions, cp;
$CW$ = Water compressibility, vol/vol/psi;
$XNPB$ = Cumulative oil production to bubble point pressure, STB;
$PB$ = Bubble point pressure, psi;
$RSI$ = Initial solution gas-oil ratio, SCF/STB;
$BOB$ = Oil formation volume factor at bubble point pressure, bbl/STB;
$RW$ = Radius of the oil-bearing portion of the reservoir, ft;
$RERW$ = Ratio of the aquifer radius to the reservoir radius; it is used only for identification purposes and not for any computations;
$C1(I)$ = Time, day;
$C2(I)$ = Average reservoir pressure, psi;
$C10(I)$ = Average pressure at oil-water contact, psi;
$C3(I)$ = Cumulative oil production, STB;
$C4(I)$ = Cumulative oil production from the bubble point, STB;
$C6(I)$ = Cumulative gas-oil ratio, SCF/STB;
$C8(I)$ = Cumulative water production, bbl;
$C9(I)$ = Cumulative water production from the bubble point, STB;
$C5(I)$ = Two-phase formation volume factor;
$C7(I)$ = Gas formation volume factor $\times 10^3$, bbl/MSCF;
$NN$ = Number of $Q_T$ and $T_D$ values provided;
$TD(I)$ = Dimensionless time function (from published tables);
$QT(I)$ = Dimensionless water influx response function (from published tables).

### EXAMPLE PROBLEM

Initial oil in place is determined from the production, rock, and fluid data for an initially undersaturated reservoir given in table 7.1. Additional information is also available, as follows:

Average porosity = 20.9%,
Oil formation volume factor at bubble point = 1.53846 res. bbl/STB,
Average aquifer permeability = 275 md,
Viscosity of water under reservoir condition = 0.25 cp,
Water compressibility = $6.8 \times 10$,
Radius of the reservoir (oil band) = 5452 ft,
Cumulative oil production to bubble point = 171.884 MSTB,

Table 7.1 Basic Performance and Fluid Property Data for an Initially Undersaturated Reservoir

| Time (days) | Average Reservoir Pressure (psi) | Pressure at Oil/Water Contact (psi) | Cumulative Oil Production (STB) | Cumulative Oil from Bubble Point Pressure (bbl/STB) | Two-Phase Formation Volume Factor (bbl/STB) | Cumulative Gas-Oil Ratio (SCF/STB) | Gas Formation Volume Factor | Cumulative Water Production (bbl) | Cumulative Water from Bubble Point Pressure (bbl/STB) |
|---|---|---|---|---|---|---|---|---|---|
| 0 | 3793 | 3793 | 0 | — | 1.5357850 | 900 | .8366010 | 0 | 0 |
| 91 | 3786 | 3788 | 13549 | — | 1.5357862 | 900 | .8366066 | 0 | 0 |
| 182 | 3768 | 3774 | 49005 | — | 1.5362877 | 900 | .8404456 | 370 | 0 |
| 273 | 3739 | 3748 | 99774 | — | 1.5370954 | 900 | .8456424 | 1030 | 0 |
| 365 | 3699 | 3709 | 171884 | — | 1.5384600 | 900 | .8532299 | 1750 | 0 |
| 456 | 3657 | 3680 | 324843 | 152959 | 1.5426846 | 900 | .8614110 | 2834 | 1084 |
| 547 | 3613 | 3643 | 528068 | 356184 | 1.5484913 | 919 | .8702281 | 4840 | 3090 |
| 638 | 3558 | 3595 | 788009 | 616125 | 1.5560508 | 914 | .8816134 | 7749 | 5999 |
| 730 | 3511 | 3547 | 1066911 | 895027 | 1.5627831 | 910 | .8916796 | 13895 | 12145 |
| 821 | 3476 | 3518 | 1339902 | 1168018 | 1.5679692 | 911 | .8993873 | 24808 | 23058 |
| 912 | 3444 | 3485 | 1615461 | 1443577 | 1.5728430 | 917 | .9065961 | 37653 | 35903 |
| 1003 | 3408 | 3437 | 1890560 | 1718676 | 1.5784815 | 937 | .9149005 | 58449 | 56699 |
| 1095 | 3375 | 3416 | 2171963 | 2000079 | 1.5838030 | 952 | .9226956 | 111863 | 110113 |
| 1186 | 3333 | 3379 | 2441226 | 2269342 | 1.5907907 | 970 | .9328846 | 163250 | 161500 |
| 1277 | 3309 | 3358 | 2713986 | 2542102 | 1.5948969 | 987 | .9388428 | 219848 | 218098 |
| 1368 | 3293 | 3338 | 2970088 | 2798204 | 1.5976815 | 1006 | .9428174 | 301256 | 297506 |
| 1460 | 3277 | 3329 | 3175948 | 3004064 | 1.6005046 | 1016 | .9469482 | 381548 | 379798 |

Cumulative water production to bubble point = 1.750 Mbbl,
Bubble point pressure = 3699 psi,
Initial solution gas-oil ratio = 900 SCF/STB.

The extent of the aquifer is 400 times the size of the pool—that is, $r_e/r_w = 20$.

# INPUT LISTING

Column legend (top of form):

| 0 = ZERO | 1 = ONE | 2 = TWO |
|---|---|---|
| Ø = ALPHA O | I = ALPHA I | Z = ALPHA Z |

Columns: STATEMENT NUMBER (1–5) · statement (7–72) · IDENTIFICATION (73–80)

```
1  EXAMPLE PROBLEM 6   NEW YORK NY  1969
2  ASSUMED INFINITE AQUIFER
3  7
4  0.-12.0.9.0E+0.0.0.-27.50E+0.3.0.-2.50.0E+0.0 -.0000068
5     1.7 8.4.-3.6.9.9.-9.0.--1.-5.3.8.4.6  5.4.5.2.-2.0.-
```

| Stmt | data | | ident |
|---|---|---|---|
| 6-1 | 3.79.3.-3.7.9.3.- | 0.- | 0.- | 9.0.- | 0.- | 0.-11.-5.3.5.7.8.5.0.- | 8.3.6.0.1.0 |
| 6-2 | 3.78.6.-3.7.8.8.- | 1.3.5.1.9.- | 0.- | 9.0.- | 0.- | 0.-11.-5.3.5.7.8.6.1.2.- | 8.3.6.0.6.6 |
| 6-3 | 3.76.8.-3.7.7.4.- | 4.9.0.0.5.- | 0.- | 9.0.- | 3.7.0.- | 0.-11.-5.3.7.0.9.5.4.- | 8.4.0.4.5.6 |
| 6-4 | 3.71.3.-3.7.4.8.- | 9.9.7.7.4.- | 0.- | 9.0.- | 1.0.3.0.- | 0.-11.-5.3.7.0.9.5.4.- | 8.4.5.6.2.4 |
| 6-5 | 3.69.9.-3.7.0.9.- | 1.7.1.8.8.4.- | 0.- | 9.0.- | 1.5.0.- | 0.-11.-5.3.8.4.6.0.0.- | 8.5.3.2.9.9 |
| 6-6 | 3.65.7.-3.6.8.0.- | 3.2.4.8.4.3.- | 1.5.2.9.5.9.- | 9.0.0.- | 2.8.3.4.- | 1.0.8.4.-11.-5.4.1.6.8.4.1.6 | 8.6.4.4.1.6 |
| 6-7 | 3.61.3.-3.6.4.5.- | 5.2.8.0.6.8.- | 3.5.6.1.1.8.4.- | 9.1.9.- | 4.8.4.0.- | 3.0.9.0.-11.-5.4.1.8.4.9.2.3.- | 8.7.0.2.2.8 |
| 6-8 | 3.55.8.-3.5.9.5.- | 7.8.8.0.0.9.- | 6.1.6.1.2.5.- | 9.1.4.- | 7.1.7.4.9.- | 5.9.9.9.-11.-5.6.1.0.5.0.8.- | 8.8.1.6.1.3.4 |
| 6-9 | 3.51.1.-3.5.4.7.- | 1.0.6.6.9.1.1.- | 8.9.5.0.2.7.- | 9.1.0.- | 1.3.8.9.5.- | 1.2.1.4.5.-11.-5.6.2.7.8.3.1.- | 8.9.1.6.7.9.6 |
| 6-10 | 3.47.6.-3.5.1.8.- | 1.3.3.9.9.0.2.- | 1.1.6.8.0.1.8.- | 9.1.1.- | 1.2.4.8.0.8.- | 2.3.0.5.8.-11.-5.6.1.7.9.6.9.2 | 8.9.3.8.7.3 |
| 6-11 | 3.44.4.-3.4.8.5.- | 1.6.1.1.5.4.6.- | 1.4.4.3.5.7.1.7.- | 9.1.7.- | 1.3.7.1.6.5.3.- | 3.1.5.9.1.0.3.-11.-5.7.2.8.4.5.1.0.- | 9.0.6.5.9.6.1 |
| 6-12 | 3.40.8.-3.4.3.7.- | 1.8.9.0.5.6.6.- | 1.7.1.8.6.1.7.1.6.- | 9.3.1.7.- | 1.5.8.4.4.9.1.- | 5.6.1.6.9.9.-11.-5.7.8.4.1.1.1.5 | 9.1.4.5.6.0.5 |
| 6-13 | 3.31.7.-3.4.1.1.6.- | 2.1.7.1.9.6.3.- | 2.1.0.0.0.0.7.9.- | 9.5.2.- | 1.1.6.3.2.5.0.- | 1.1.6.1.5.0.0.-11.-5.9.0.1.7.9.6.7 | 9.1.2.6.9.5.6 |
| 6-14 | 3.39.3.-3.3.7.9.- | 2.4.1.4.1.2.2.6.- | 2.2.6.9.3.4.2.- | 9.7.0.- | 1.6.3.2.5.0.- | 2.1.8.0.9.8.-11.-5.9.4.8.7.6.9 | 9.5.2.8.8.4.6 |
| 6-15 | 3.30.9.-3.3.5.1.8.- | 2.7.1.3.9.8.6.- | 2.5.4.2.1.0.2.- | 9.8.1.7.- | 2.1.9.8.4.8.- | 2.5.4.2.1.0.2.-11.-5.9.4.8.7.6.9 | 9.5.8.8.4.2.8 |
| 6-16 | 3.29.3.-3.3.3.8.- | 2.9.7.0.0.8.8.- | 2.7.9.8.2.0.4.- | 10.0.6.- | 1.3.0.1.3.5.6.- | 2.9.9.5.0.6.-11.-5.9.7.6.8.1.1.5 | 9.4.2.8.7.1.4 |

| STATEMENT NUMBER | | |
|---|---|---|
| 7 | 2.4 | |
| 7.1 | .1.0.0.0.0 | E.+.0.0.-.1.0.0.0.0.1.0.0.E.+.0.1.0 |
| 7.2 | .1.1.5.0.0 | E.+.0.2.-.9.9.4.9.0.9.E.+.0.1.1 |
| 7.3 | .1.3.0.0.0 | E.+.0.2.-.1.1.6.1.7.4.2.6.E.+.0.1.2 |
| 7.4 | .1.4.5.0.0.0 | E.+.0.2.-.1.2.1.8.9.7.1.E.+.0.1.2 |
| 7.5 | .1.6.0.0.0.0 | E.+.0.2.-.1.2.8.6.9.1.1.E.+.0.1.2 |
| 7.6 | .1.7.5.0.0 | E.+.0.2.-.1.3.4.2.4.7.1.E.+.0.2 |
| 7.7 | .1.9.0.0.0.0 | E.+.0.2.-.3.9.6.2.1.6.E.+.0.1.2 |
| 7.8 | .1.1.0.5.0.0 | E.+.0.3.-.1.4.4.8.5.8.E.+.0.1.2 |
| 7.9 | .1.1.2.0.1.0.0 | E.+.0.3.-.1.4.9.1.9.6.8.E.+.0.1.2 |
| 7.10 | .1.1.3.5.0.0 | E.+.0.3.-.1.5.4.9.7.1.6.E.+.0.1.2 |
| 7.11 | .1.1.5.0.0.0 | E.+.0.3.-.1.5.9.8.1.9.5.E.+.0.1.2 |
| 7.12 | .1.1.6.5.0.0 | E.+.0.3.-.1.6.4.7.3.7.E.+.0.1.2 |
| 7.13 | .1.1.8.0.0.0 | E.+.0.3.-.1.6.9.5.1.1.2.E.+.0.1.2 |
| 7.14 | .1.1.9.5.0.0 | E.+.0.3.-.1.7.4.2.6.6.E.+.0.1.2 |
| 7.15 | .2.1.1.0.0.0 | E.+.0.3.-.1.7.8.1.8.8.6.E.+.0.1.2 |
| 7.16 | .2.2.5.6.6 | E.+.0.3.-.1.8.3.4.9.1.7.E.+.0.1.2 |
| 7.17 | .2.6.5.6.6 | E.+.0.3.-.1.9.5.5.8.8.E.+.0.2 |
| 7.18 | .2.9.0.6.0 | E.+.0.3.-.1.0.3.0.1.2.E.+.0.5 |
| 7.19 | .3.1.5.0.0 | E.+.0.3.-.1.1.0.3.7.E.+.0.3 |
| 7.26 | .4.5.5.0.0 | E.+.0.3.-.1.1.5.0.2.5.E.+.0.3 |
| 7.21 | .4.8.0.0.0 | E.+.0.3.-.1.1.5.7.1.8.E.+.0.3 |
| 2.22 | .5.1.0.0.0 | E.+.0.3.-.1.1.6.5.4.4.5.E.+.0.3 |
| 7.23 | .5.5.0.0.0 | E.+.6.3.-.1.7.1.6.3.6.E.+.0.3 |
| 7.24 | .5.8.0.0.0 | E.+.0.3.-.1.8.1.4.7.E.+.0.3 |

## SOURCE LISTING

```
C          THIS PROGRAM DETERMINES THE ORIGINAL-OIL-IN-PLACE BY MATERIAL
C          BALANCE METHOD FOR RESERVOIRS WITH PARTIAL WATER-DRIVE(UNSTEADY)
C          WITH NO INITIAL GAS CAP.
C          THE RATIO OF RESERVOIR RADIUS TO AQUIFER RADIUS VALUE AND CORRESPOND
C          ING DIMENSIONLESS TIME AND 'QT' FUNCTION RELATIONSHIP IS PROVIDED
C          BY THE PROGRAMMER.
C          MATERIAL BALANCE CALCULATIONS BELOW BUBBLE-POINT PRESSURE ARE
C          PERFORMED FOR STABILITY OF CALCULATIONS.  CALCULATIONS ABOVE
C          BUBBLE-POINT AND NEAR THE BUBBLE POINT ARE FULLY SENSITIVE TO THE
C          RESERVOIR PRESSURE.  VERY ACCURATE RESERVOIR  PRESSURE
C          DATA IS NOT AVAILABLE (IN GENERAL) FOR MOST RESERVOIR STUDY

           DIMENSION C1(20),C2(20),C3(20),C4(20),C5(20),C6(20),C7(20),C8(20)
          *,C9(20),C10(20),C11(20),C12(20),C13(20),C14(20),C15(20),C16(20),
          *C17(20),C18(20),C19(20),C20(20),C21(20),C22(20),C23(20),C24(20),
          *C25(20),C26(20),C27(20),C28(20),C29(20),C30(20),C31(20),C32(20)
          *,C33(20),C34(20),C35(20),C36(20),C37(20),C38(20),C39(20),C40(20)
          *,C41(20),C42(20),C43(20),C44(20),C45(20),TD(50),CT(50),A(20),B(20)
          *,QT(50),YCAL(20),YOBS(20)
C
           READ(5,2)(A(I),I=1,20)
           READ(5,2)(B(I),I=1,20)
        2  FORMAT(20A4)
C          A AND B ARE ALPHANUMERIC VECTORS FOR PROBLEM IDENTIFICATION AND
C          HEADING PURPOSES.
           WRITE(6,3)(A(I),I=1,20)
           WRITE(6,3)(B(I),I=1,20)
        3  FORMAT(//,3X,20A4)
           WRITE(6,1)
        1  FORMAT(/,3X,'DETERMINATION OF ORIGINAL OIL IN PLACE')
           WRITE(6,11)
       11  FORMAT(3X,'UNSTEADY STATE MATERIAL BALANCE METHOD FOR A'
          *,/3X,'RESERVOIR WITH PARTIAL WATER DRIVE')
C          N=NUMBER OF PRESSURE-PRODUCTION DATA POINTS TO BE UTILIZED,
C          INCLUDING INITIAL RESERVOIR PRESSURE POINT
C          APOR=AVG. POROSITY, FRACTION
C          APERMW=AVG. PERMEABILITY TO WATER AT RESIDUAL OIL SATURATION, MD
C          AVISW=WATER VISCOSITY AT RESERVOIR CONDITION, CP
C          CW=WATER COMPRESSIBILITY, VOL/VOL/PSI
C          XNPB=CUMULATIVE OIL PRODUCTION TO BUBBLE POINT PRESSURE,
C          BBL(ST.TK)
C          WPB=CUMULATIVE WATER PRODUCTION TO BUBBLE POINT PRESSURE, BBL
C          PB=BUBBLE POINT PRESSURE, PSIA
C          PSI=INITIAL SOLN GAS OIL RATIO, SCF/STB
C          BOB=FVF AT BUBBLE POINT PRESSURE, RESERVOIR BBL/STB
C          RW=RADIUS OF THE RESERVOIR(OIL BEARING PORTION OF THE STRUCTURE), FT
C          RERW=RATIO OF RESERVOIR RADIUS TO AQUIFER RADIUS. DIMENSIONLESS
C          TIME VS. QT FUNCTION VALUES FOR THE SPECIFIC RERW VALUE MUST BE
C          PROVIDED BY THE USERS.
C          C1=TIME, DAYS
C          C2=AVERAGE RESERVOIR PRESSURE (STATIC), PSIA
C          C4=CUMULATIVE OIL PRODUCTION FROM BUBBLE POINT PRESSURE,MUST BE
C          EQUAL TO ZERO AT OR ABOVE ABOVE BUBBLE POINT PRESSURE, STB
C          C5=TWO-PHASE FORMATION VOLUME FACTOR, RES. BBL/STB
C          C6=CUMULATIVE GAS-OIL RATIO,IE, PRODUCED GAS/PRODUCED OIL, SCF/STB
C          C7=GAS FORMATION VOLUME FACTOR, RES. BBL/SCF
C          C8=CUMULATIVE WATER PRODUCTION, BBL
C          C9=CUMULATIVE WATER PRODUCTION BELOW BUBBLE POINT, BBL MUST BE
C          ZERO AT OR ABOVE BUBBLE POINT
C          C10=AQUIFER PRESSURE OR PRESSURE AT OIL WATER CONTACT, PSIA
           READ(5,4)N
        4  FORMAT(I2)
           READ(5,5)APOR,APERMW,AVISW,CW
           READ(5,8)XNPB,PB,RSI,BOB,RW,RERW
        5  FORMAT(3E10.4,F10.7)
        8  FORMAT(F10.0,2F5.0,F7.5,F8.0,F3.0)
           WRITE(6,6)RERW
        6  FORMAT(//,3X,'ASSUMED RATIO OF THE RESERVOIR TO THE',
          X' RADIUS OF THE AQUIFER',3X,F10.2)
           DO 10 I=1,N
           READ(5,7)C1(I),C2(I),C10(I),C3(I),C4(I),C6(I),C8(I),C9(I),C5(I),
          *C7(I)
           C7(I)=C7(I)*.001
        7  FORMAT(3F6.0,2F10.0,F7.0,2F7.0,F9.7,F9.7)
       10  CONTINUE
C
C          NN=NUMBER OF TD-QT VALUES TO BE READ
           READ(5,15)NN
       15  FORMAT(I3)
C
```

```
C         TD-DIMENSIONLESS TIME VALUES MUST BE READ IN INCREASING ORDER
          DO 20 I=1,NN
          READ(5,9)TD(I),QT(I)
    9     FORMAT(2E10.5)
   20     CONTINUE
          WRITE(6,600)
  600     FORMAT(/,3X,'PRODUCTION, PRESSURE AND FLUID DATA PROVIDED'/)
          WRITE(6,610)
  610     FORMAT(3X,'TIME',2X,'AVE. RES',3X,'AVE. AQ',3X,'CUM. OIL PROD',
         *3X,'CUM OIL FROM',3X,'CUM WATER',3X,'T-P   ',3X,'CUM  ',/,11X,
         *'PRES',4X,'PRES.',22X,'B.P.PRES',8X,'PROD',6X,'FVF',5X,'GOR  '
         *,/,3X,'DAYS',4X,'PSI',5X,'PSI',14X,'STB',10X,'STB',11X,'STB',
         *16X,'CF/BBL')
          DO 650 I=1,N
          WRITE(6,620)C1(I),C2(I),C10(I),C3(I),C4(I),C8(I),C5(I),C6(I)
  620     FORMAT(2X,F5.0,3X,F6.0,3X,F6.0,3X,F12.0,3X,F12.0,3X,F9.0,
         *3X,F5.3,3X,F7.2)
  650     CONTINUE
          DO 700 I=1,N
          IF(C2(I).LT.PB) GO TO 710
          IF(C2(I).GE.PB)JJ=0
  700     CONTINUE
          GO TO 711
  710     JJ=I
  711     CONTINUE
          IF(JJ.EQ.0) GO TO 464
          T=0.006323*APERMW/(APOR*AVISW*CW*RW**2)
          C18(1)=0.
          C19(1)=0.
          DO 25 I=2,N
          C18(I)=T*C1(I)
          CALL SPLINE(TD,QT,NN)
          CALL GRATA(C18(I),AA,NN)
          CALL TERPA(C18(I),TX,NN)
          C19(I)=TX
   25     CONTINUE
          C16(1)=0.
          C21(JJ-1)=0.
          C30(JJ-1)=0.
          C28(JJ-1)=0.
          C24(JJ-1)=0.
          C26(JJ-1)=0.
          C23(JJ-1)=0.
          C35(JJ-1)=0.
          DO 100 I=2,N
          C11(I)=C6(I)-RSI
          C12(I)=C7(I)*C11(I)
          C13(I)=C5(I)+C12(I)
          C14(I)=C13(I)*C4(I)
          C15(I)=C14(I)+C9(I)
          C16(I)=C10(I-1)-C10(I)
          C17(I)=0.5*(C16(I-1)+C16(I))
  100     CONTINUE
          DO 200 I=2,N
          SUM=0.0
          DO 250 J=2,I
          SUM=SUM+C17(J)*C19(I+2-J)
  250     CONTINUE
          C20(I)=SUM
  200     CONTINUE
          DO 350 I=JJ,N
          C21(I)=C20(I)-C20(JJ-1)
  350     CONTINUE
          DO 400 I=JJ,N
          C22(I)=C5(I)-BOB
          C23(I)=C15(I)/C22(I)
          C24(I)=C24(I-1)+C23(I)
          C25(I)=C21(I)/C22(I)
          C26(I)=C26(I-1)+C25(I)
          C27(I)=C23(I)*C25(I)
          C28(I)=C28(I-1)+C27(I)
          C29(I)=C25(I)**2
          C30(I)=C30(I-1)+C29(I)
          C31(I)=C26(I)**2
          C32(I)=C30(I)*C24(I)
          C33(I)=C26(I)*C28(I)
          C34(I)=C32(I)-C33(I)
          C35(I)=C35(I-1)+1.
          C36(I)=C35(I)*C30(I)
          C37(I)=C36(I)-C31(I)
          IF(C37(I).LE.0.)GO TO 399
          IF(C37(I).GT.0.)C38(I)=C34(I)/C37(I)
          C39(I)=C35(I)*C38(I)
          C40(I)=C24(I)-C39(I)
          IF(C26(I).GT.0.)C41(I)=C40(I)/C26(I)
          IF(C26(I).LE.0.)C41(I)=0.
          GO TO 400
```

```
399 CONTINUE
    C38(I)=0.
400 CONTINUE
    BAVG=(C41(N-3)+C41(N-2)+C41(N-1)+C41(N))*0.25
    AND=(C38(N-3)+C38(N-2)+C38(N-1)+C38(N))*0.25
    OIP=AND+XNPB
440 CONTINUE
    WRITE(6,451)
451 FORMAT(/10X,'ORIGINAL OIL-IN-PLACE CALCULATIONS')
    WRITE(6,450)
450 FORMAT(//,3X,' TIME',3X,'RES.PRES.',3X,'OIP AT BP',3X,'PROP.FAC',
   */,4X,'DAYS',5X,'PSIA',9X,'STB',9X,'Y')
    DO 460 I=JJ,N
    WRITE(6,461)C1(I),C2(I),C38(I),C41(I)
461 FORMAT(4X,F5.0,3X,F9.0,3X,F9.0,3X,F8.1)
460 CONTINUE
    WRITE(6,462) OIP
462 FORMAT(/,3X,'ORIGINAL OIL IN PLACE',3X,F10.0,3X,'STB')
    GO TO 465
464 CONTINUE
    WRITE(6,466)
465 FORMAT(3X,'RESERVOIR PRESSURE ALWAYS ABOVE BUBBLE POINT,'
   *,' MB CALCULATIONS ARE NOT POSSIBLE IN THIS PROGRAM.')
465 CONTINUE
    STOP
    END
    SUBROUTINE SPLINE(XI,YI,N)
    DIMENSION W(99),Q(99),B(99),A(99),C(99),S(99),Z(99)
    DIMENSION XI(50),YI(50),X(99),Y(99)
    DATA A(1),C(1),Z(1)/-1.0,0.,0./
    DO 11 I=1,N
    X(I)=XI(I)
    Y(I)=YI(I)
11  CONTINUE
    DO 1 J=2,N
    W(J)=X(J)-X(J-1)
1   CONTINUE
    NM=N-1
    DO 2 J=2,NM
    WJ=W(J)
160 WP=W(J+1)
    WS=WJ+WP
    QJ=WJ/WS
    Q(J)=QJ
    QA=1.-.5*QJ*A(J-1)
    A(J)=.5*(1.-QJ)/QA
    B(J)=3.*(WP*Y(J-1)-WS*Y(J)+WJ*Y(J+1))/WP/WJ/WS
    C(J)=(B(J)-.5*QJ*C(J-1))/QA
2   CONTINUE
    S(N)=C(NM)/(1.+A(NM))
    S(NM)=S(N)
    NMM=N-2
    DO 3 JJ=1,NMM
    J=NMM-JJ+1
    S(J)=C(J)-A(J)*S(J+1)
3   CONTINUE
    DO 4 J=2,N
    Z(J)=Z(J-1)+.5*W(J)*(Y(J)+Y(J-1)-.0825*W(J)**2*(S(J)+S(J-1)))
4   CONTINUE
    RETURN
C
    ENTRY DERIVA(XV,YV,N)
    DO 5 JJ=2,N
    J=JJ
    IF(XV.GT.X(J))GOTO 5
    GOTO 6
5   CONTINUE
6   WJ=W(J)
    D1=(XV-X(J-1))/WJ
    D2=(X(J)-XV)/WJ
    YV=(Y(J)-Y(J-1))/WJ+.5*WJ*((D1*D1-.333333)*S(J)
   *-(D2*D2-.333333)*S(J-1))
    RETURN
    ENTRY TERPA(XV,YV,N)
    DO 7 JJ=2,N
    J=JJ
    IF(XV.GT.X(J)) GOTO 7
    GOTO 8
7   CONTINUE
8   WJ=W(J)
    D1=(XV-X(J-1))/WJ
    D2=(X(J)-XV)/WJ
    D3=WJ*WJ/6.
    YV=D1*(Y(J)+D3*(D1*D1-1.)*S(J))
    YV=YV+D2*(Y(J-1)+D3*(D2*D2-1.)*S(J-1))
    RETURN
    ENTRY GRATA(XV,YV,N)
    DO 9 JJ=2,N
```

```
      J=JJ
      IF(XV.GT.X(J)) GOTO 9
      GOTO 10
    9 CONTINUE
   10 WJ=W(J)
      D1=((XV-X(J-1))/WJ)**2
      D2=((X(J)-XV)/WJ)**2
      D3=.0825*WJ*WJ
      YV=Z(J-1)+.5*WJ*(D1*(Y(J)+D3*(D1-2.)*S(J))+(1.-D2)*(Y(J-1)+D3*(D2
     *-1.)*S(J-1)))
      RETURN
      END
//GO.SYSIN DD *
EXAMPLE PROBLEM 6  NEWYCRK   NY    P69
ASSUMED INFINITE AQUIFER
17
      .2090  275.            .2500   .0000068
    171884.3699. 900.1.53846    5452.20.00
0.       3793. 3793.  0.         0.        900. 0.       0.       1.53578500.8366010
91.      3786. 3788.  13549.     0.        900. 0.       0.       1.53578620.8366066
182.     3768. 3774.  49005.     0.        900. 370.     0.       1.53628770.8404456
273.     3739. 3748.  99774.     0.        900. 1030.    0.       1.53709540.8456424
365.     3699. 3709.  171884.    0.        900. 1750.    0.       1.53846000.8532299
456.     3657. 3680.  324843.    152959.   900. 2834.    1084.    1.54268460.8614110
547.     3613. 3643.  528068.    356184.   919. 4840.    3090.    1.54849230.8702281
638.     3558. 3595.  788009.    616125.   914. 7749.    5999.    1.55605080.8816134
730.     3511. 3547.  1066911.   895027.   910. 13895.   12145.   1.56278310.8916796
821.     3476. 3518.  1339902.   1168018.  911. 24808.   23058.   1.56796920.8993873
912.     3444. 3485.  1615461.   1443577.  917. 37653.   35903.   1.57284300.9065961
1003.    3408. 3437.  1890560.   1718676.  937. 58449.   56699.   1.57848150.9145005
1095.    3376. 3416.  2171963.   2000079.  952. 111863.110113.1.58380300.9226956
1186.    3333. 3379.  2441226.   2269342.  970. 163250.161500.1.59079073.9328846
1277.    3309. 3358.  2713986.   2542102.  987. 219848.218098.1.59489690.9388428
1368.    3293. 3338.  2970088.   2798204.  1006.301256.299506.1.59768150.9428714
1460.    3277. 3329.  3175948.   3004064.  1016.381548.379798.1.60050460.9469482
   24
0.       0.
.15000E 02.99490E 01
.30000E 02.16742E 02
.45000E 02.22897E 02
.60000E 02.28691E 02
.75000E 02.34247E 02
.90000E 02.39626E 02
.10500E 03.44858E 02
.12000E 03.49968E 02
.13500E 03.54976E 02
.15000E 03.59895E 02
.16500E 03.64737E 02
.18000E 03.69512E 02
.19500E 03.74226E 02
.21000E 03.78886E 02
.22500E 03.83497E 02
.26500E 03.95588E 02
.29000E 03.10302E 03
.31500E 03.11037E 03
.45500E 03.15025E 03
.48000E 03.15718E 03
.51000E 03.16544E 03
.55000E 03.17636E 03
.58000E 03.18447E 03
/*
//
```

# OUTPUT LISTING

EXAMPLE PROBLEM 6   NEWYORK   NY   P69

ASSUMED INFINITE AQUIFER

DETERMINATION OF ORIGINAL OIL IN PLACE
UNSTEADY STATE MATERIAL BALANCE METHOD FOR A
RESERVOIR WITH PARTIAL WATER DRIVE

ASSUMED RATIO OF THE RESERVOIR TO THE RADIUS OF THE AQUIFER   20.00

PRODUCTION, PRESSURE AND FLUID DATA PROVIDED

| TIME DAYS | AVE. RES PRES. PSI | AVE. AQ PRES. PSI | CUM. OIL PROD STB | CUM OIL FROM B.P.PRES STB | CUM WATER PROD. STB | T-P FVF | CUM GOR CF/BBL |
|---|---|---|---|---|---|---|---|
| 0. | 3793. | 3793. | 0. | 0. | 0. | 1.536 | 900.00 |
| 91. | 3786. | 3788. | 13549. | 0. | 0. | 1.536 | 900.00 |
| 182. | 3768. | 3774. | 49005. | 0. | 370. | 1.536 | 900.00 |
| 273. | 3739. | 3748. | 99774. | 0. | 1020. | 1.537 | 900.00 |
| 365. | 3699. | 3709. | 171884. | 0. | 1750. | 1.538 | 900.00 |
| 456. | 3657. | 3680. | 324843. | 152959. | 2824. | 1.543 | 900.00 |
| 547. | 3613. | 3643. | 528068. | 356184. | 4840. | 1.548 | 914.00 |
| 638. | 3558. | 3595. | 788309. | 616125. | 7749. | 1.556 | 914.00 |
| 730. | 3511. | 3547. | 1066911. | 895027. | 13855. | 1.563 | 910.00 |
| 821. | 3476. | 3518. | 1339902. | 1168018. | 24808. | 1.568 | 911.00 |
| 912. | 3444. | 3485. | 1615461. | 1443577. | 37653. | 1.573 | 917.00 |
| 1003. | 3408. | 3437. | 1890560. | 1718676. | 58449. | 1.578 | 937.00 |
| 1095. | 3376. | 3416. | 2171963. | 2000079. | 111863. | 1.584 | 952.00 |
| 1186. | 3333. | 3379. | 2441226. | 2269342. | 163250. | 1.591 | 970.00 |
| 1277. | 3309. | 3358. | 2713986. | 2542102. | 219848. | 1.595 | 987.00 |
| 1368. | 3293. | 3338. | 2970088. | 2798204. | 301256. | 1.598 | 1006.00 |
| 1460. | 3277. | 3329. | 3175948. | 3004064. | 381548. | 1.601 | 1016.00 |

ORIGINAL OIL-IN-PLACE CALCULATIONS

| TIME DAYS | RES.PRES. PSIA | OIF AT BP STB | PROP.FAC Y |
|---|---|---|---|
| 456. | 3657. | 5142048. | 0.0 |
| 547. | 3613. | 45625728. | 26.4 |
| 638. | 3558. | 45625728. | 59.3 |
| 730. | 3511. | 30174304. | 151.1 |
| 821. | 3476. | 22887504. | 193.4 |
| 912. | 3444. | 22402704. | 196.1 |
| 1003. | 3408. | 23384496. | 190.7 |
| 1095. | 3376. | 23584336. | 189.6 |
| 1186. | 3333. | 24458512. | 184.9 |
| 1277. | 3309. | 24775360. | 183.3 |
| 1368. | 3293. | 24445088. | 184.9 |
| 1460. | 3277. | 24761696. | 183.4 |

ORIGINAL OIL IN PLACE   24782032.   STB

**BIBLIOGRAPHY**

Craft, B. C., and Hawkins, M. F.: *Applied Petroleum Reservoir Engineering,* Prentice-Hall, Englewood Cliffs, N.J. (1959).

Guerrero, E. T.: *Practical Reservoir Engineering,* The Petroleum Publishing Company, Tulsa (1968).

# II    FUTURE RESERVOIR PERFORMANCE (OIL)

# 8    PERFORMANCE PREDICTION BY PRODUCTION DECLINE ANALYSIS

**PURPOSE**

Extrapolation of characteristic production decline trends from past performance is used to predict future performance of a producing well or reservoir. The historical data must represent unrestricted (capacity) production. The basis of extrapolation is the assumption that the future behavior of a well/reservoir will be governed by the past trend or mathematical relation that represents its past performance.

Regression analysis technique is applied to determine the optimal values of the characteristic parameters (decline constant and exponent) of the generalized decline curve equation.

Use of this program will eliminate extensive data plotting. It is also believed that this method will provide better results than manual extrapolation of the past trend.

**METHOD**

Rate-time relationship:

$$q = q_o\left(1 + \frac{b_t}{a_i}\right)^{-1/b}. \tag{8.1}$$

Rate-cumulative relationship:

$$Q = \frac{q_o a_o}{1 - b}(q_o^{1-b} - q^{1-b}), \tag{8.2}$$

where

$q$ = Production rate at time t;
$q_o$ = Initial production rate;
$b$ = Decline exponent, a nonnegative constant $0 \leq b \leq 3$;
$a_o$ = Decline constant, a nonnegative constant $0 \leq a_o \leq 0.999$;
$t$ = time;
$Q$ = Cumulative production between the rates $q_o$ and $q$.

Equations (8.1) and (8.2) are generalized decline equations. The special cases are exponential and harmonic declines, where the decline exponent takes on the value of 0 and 1 respectively.

The method employs a least-squares function and actual production data over a period of time to obtain the best fit of the data in the form of the generalized decline equation (8.1) by optimizing the values of $a_o$ and $b$.

**PROGRAM DESCRIPTION**

The program is designed to handle both gas or oil production data. Units of production rate and time input are bbl/mo or MCF/mo and month. In the prediction mode, annual production and end of the year rate are computed for the number of years desired or to the economic limit rate of production, whichever comes first. Up to 20 years of predictions can be performed. The input data set may contain 60 months of data.

The least-squares optimization code uses a BSOLVE algoritnm based on Marquardt's method (developed by W. E. Ball; see Ball 1973).

**Input Format**

| Card | Format | Content |
|------|--------|---------|
| 1 | 20A4 | ANAME(I) |
| 2 | 20A4 | BNAME(I) |
| 3 | I3, I2 | NN, IOPT |
| 4 | F3.0, F5.0 | MO, YR |
| 5–NN | F3.0, F9.0, F10.0 | X(JJ), Y(JJ), ANP(JJ) |
| 6 | I3, F5.0 | NMP, ECL |

**Description of Input Parameters**

ANAME(I) = Uses up to 80 alphanumeric characters for project identification;
BNAME(I) = Uses up to 80 alphanumeric characters for project identification;

NN = Number of months for which historical data are available;
IOPT = Production stream option card—i.e., gas or oil: IOPT = 1, oil stream; IOPT = 2, gas stream;
MO = Month of starting historical data;
YR = Year of starting historical data;
$X(JJ)$ = Month (1, 2, 3, etc.);
$Y(JJ)$ = Corresponding monthly production, bbl/mo or MCF/mo;
$ANP(JJ)$ = Cumulative production from the starting point, bbl/mo or MCF/mo;
NMP = Number of years for which prediction is desired;
ECL = Economic limit rate of production, bbl/mo or MCF/mo.

## EXAMPLE PROBLEM

The monthly lease production data from January 1952 to December 1952 are given in table 8.1. Determine future production from the lease for the next five years or to economic limit rate of 60 bbl/mo, whichever comes first.

Table 8.1 Monthly Production Data

| Month | Year | Monthly Production (bbl) | Cumulative Production (bbl) |
|---|---|---|---|
| 1 | 1952 | 29500 | 29500 |
| 2 | 1952 | 23606 | 53106 |
| 3 | 1952 | 21301 | 74407 |
| 4 | 1952 | 19318 | 93725 |
| 5 | 1952 | 17599 | 111324 |
| 6 | 1952 | 16100 | 127424 |
| 7 | 1952 | 14785 | 142209 |
| 8 | 1952 | 13624 | 155833 |
| 9 | 1952 | 12595 | 168428 |
| 10 | 1952 | 11678 | 180106 |
| 11 | 1952 | 10858 | 190964 |
| 12 | 1952 | 10121 | 201085 |

**INPUT LISTING**

# FORTRAN CODING FORM

WW/CC   NAME _____   DATE _____

PROGRAM _____
ROUTINE _____

0 = ZERO   Ø = ALPHA O   1 = ONE   I = ALPHA I   2 = TWO   Z = ALPHA Z

| CARD | STATEMENT NUMBER | 6 7 8 9 ... (columns) |
|---|---|---|
| 1. | | EXAMPLE PROBLEM #1  SALTLAKE CITY, UTAH |
| 2. | | WELL KANSAS LEASE ARBUCKLE LIME PRODUCTION |
| 3. | | 12.1 |
| 4. | 1195 2. | |
| 5-1 | 1. | 29500.  295.00. |
| 5-2 | 2. | 23606.  531.06. |
| 5-3 | 3. | 21301.  744.07. |
| 5-4 | 4. | 19318.  931725. |
| 5-5 | 5. | 17599.  111324. |
| 5-6 | 6. | 16100.  127424. |
| 5-7 | 7. | 14785.  142209. |
| 5-8 | 8. | 13624.  155833. |
| 5-9 | 9. | 12595.  168428. |
| 5-10 | 10. | 11678.  180106. |
| 5-11 | 11. | 10858.  190964. |
| 5-12 | 12. | 10121.  201085. |
| 6. | 15 | 60. |

## SOURCE LISTING

```
C          THIS PROGRAM ANALYZES PRODUCTION DECLINE DATA TO CHARACTERIZE THE
C          TYPE OF DECLINE BY DETERMINING THE DECLINE COEFFICIENT ANC THE
C          DECLINE EXPONENT OF THE GENERALIZED DECLINE EQUATION. THESE VALUES
C          ARE THEN UTILIZED TO MAKE FUTURE PERFORMANCE PREDICTION AND
C          RECOVERY. METHOD INVOLVES UTILIZATION OF LEAST SQUARES NONLINEAR
C          REGRESSION TECHNIQUE.
C
           REAL MO,MOP
           DIMENSION P(100),A(5,5),AC(5,5),X(50),B(2),Z(50),Y(50),BV(2),
          *BMIN(2),BMAX(2),FV(2),DV(2),ANAME(20),BNAME(20),MOP(60),
          *Q(60),QT(160),QP(180),QTP(180),MOM(160),YOP(160),ANP(150)
C
C          B(1)=EXPONENT
C          B(2)=COEFFICIENT
           EXTERNAL FUNC
           COMMON X,Y
           READ(5,12)(ANAME(I),I=1,20)
           READ(5,12)(BNAME(I),I=1,20)
        12 FORMAT(20A4)
           WRITE(6,13)
        13 FORMAT(' DECLINE CURVE ANALYSIS AND PERFORMANCE PREDICTION')
           WRITE(6,14)(ANAME(I),I=1,20)
           WRITE(6,14)(BNAME(I),I=1,20)
        14 FORMAT(/,40X,20A4)
C          READ IN NUMBER OF DATA POINTS
           READ(5,11)NN,IOPT
        11 FORMAT(I3,I2)
           KK=2
C          BMIN(J) AND BMAX(J) ARE THE LIMITS ON THE DECLINE COEFFICIENTS AND
C          EXPONENT
           BMIN(1)=0.
           BMIN(2)=0.
           BMAX(1)=2.
           BMAX(2)=.999
C          IOPT=1     BBL/MO UNIT OF PRODUCTION
C          IOPT=2     MCF/MO UNIT OF PRODUCTION
           FNU=0.
           FLA=0.
           TAU=0.
           EPS=0.
           PHMIN=0.
           I=0
           KD=KK
           FV(1)=0.
           FV(2)=0.
C          MO,YR=MONTH AND YEAR OF STARTING HISTORICAL DATA TO BE READ IN
           READ(5,2000)MO,YR
      2000 FORMAT(F3.0,F5.0)
           DO 2001 JJ=1,NN
           READ(5,2002)X(JJ),Y(JJ),ANP(JJ)
      2002 FORMAT(F3.0,F9.0,F10.0)
C          X(JJ)=MONTH WITH THE STARTING MONTH IE. 1,2,3 ETC.
C          Y(JJ)=CORRESPONDING MONTHLY PRODUCTION, MCF OR BBL PER MONTH
      2001 CONTINUE
           DO 2006 I=1,NN
           MOP(I)=MO
           YOP(I)=YR
           MO=MO+1
           IF(MO.GT.12)YR=YR+1.
           IF(MO.GT.12)MO=1.
      2006 CONTINUE
      2007 CONTINUE
           I=0
           B(1)=0.5
           B(2)=(ABS(Y(1)-Y(2)))/(.5*(Y(1)+Y(2)))
           DO 100 J=1,KK
           BV(J)=1.
       100 CONTINUE
           ICON=KK
           ITER=0.
       200 CALL BSOLVE(KK,B,NN,Z,Y,PH,FNU,FLA,TAU,EPS,PHMIN,I,ICON,FV,DV,BV,
          1BMIN,BMAX,P,FUNC,DERIV,KD,A,AC,GAMM)
C
           ITER=ITER+1
           IF (ICON) 10, 300, 200
        10 IF (ICON+1) 20, 60, 200
        20 IF (ICON+2) 30, 70, 200
        30 IF (ICON+3) 40, 80, 200
        40 IF (ICON+4) 50, 90, 200
```

```
   50 GO TO 95
   60 WRITE (6,4)
    4 FORMAT (//,2X,'NO FUNCTION IMPROVEMENT POSSIBLE ')
      GO TO 300
   70 WRITE (6,5)
    5 FORMAT (//,2X, 'MORE UNKNOWNS THAN FUNCTIONS')
      GO TO 300
   80 WRITE (6,6)
    6 FORMAT (//,2X, 'TOTAL VARIABLES ARE ZERO')
      GC TO 300
   90 WRITE (6,7)
    7 FORMAT (//,2X,'CORRECTIONS SATISFY CONVERGENCE REQUIREMENTS BUT'
     1,' LAMDA FACTOR (FLA) STILL LARGE')
      GO TO 300
   95 WRITE (6,8)
    8 FORMAT (//,2X, 'THIS IS NOT POSSIBLE')
  300 CONTINUE
      BZERO=B(1)
      AZERO=B(2)
      WRITE(6,2010)
 2010 FORMAT(//,3X,'MONTH   YEAR    ACT.PRODUCTION    CALC.',
     *' PRODUCTION    ACT.CUM.PROD.    CALC.CUM.PRODN',3X,'COEFF',
     *'    EXPO')
      DO 2020 I=2,NN
      Q(I)=Y(1)*((1.+BZERO*AZERO*I)**(-1./BZERO))
      QT(I)=((Y(1)**BZERO)/(AZERO*(1.-BZERO)))*((Y(1)**(1.-BZERO)-
     *Y(I)**(1.-BZERO)))
      WRITE(6,2021)MOP(I),YOP(I),Y(I),Q(I),ANP(I),QT(I),AZERO,BZERO
 2021 FORMAT(F7.0,F7.0,F13.0,F18.0,F16.0,F18.0,F11.3,F7.3)
 2020 CONTINUE
 2030 FORMAT(I3,F5.0)
      READ(5,2030)NMP,ECL
    C NMP = NO. OF YEARS FOR WHICH PREDICTION IS DESIRED.
    C ECL = ECONOMIC LIMIT RATE OF PRODUCTION
      N1N = NN+NMP*12
      N2N=NN+1
      DO 2050 I=N2N,N1N
      QP(I)=Y(1)*((1.+BZERO*AZERO*I)**(-1./BZERO))
      IF(QP(I).LT.ECL)GO TO 2051
      QT(I)=((Y(1)**BZERO)/(AZERO*(1.-BZERO)))*((Y(1)**
     *(1.-BZERO))-(QP(I)**(1.-BZERO)))
 2050 CONTINUE
      GO TO 2052
 2051 CONTINUE
 2052 CONTINUE
      WRITE(6,2053)
 2053 FORMAT(///,3X,'MO ',1X,'YEAR',5X,'RATE',6X,'YEARLY PROD.'
     *,3X,'CUM. PROD.')
      IF(IOPT.EQ.1) WRITE(6,2054)
      IF(IOPT.EQ.2) WRITE(6,2055)
 2054 FORMAT(15X,'BO/M',10X,'BBL',11X,'BBL')
 2055 FORMAT(15X,'MCF.M',9X,'MCF',11X,'MCF')
      N2N=NN+12
      YR=YOP(NN)
      J=0
      DO 2056 I=N2N,N1N,12
      J=J+1
      YOP(I)=YR+1.
      MOP(I)=I-J*12
      ANP(I)=QT(I)-QT(I-12)
      YR=YOP(I)
      WRITE(6,2057)MOP(I),YOP(I),QP(I),ANP(I),QT(I)
 2057 FORMAT(2X,F3.0,F6.0,F8.0,F17.0,F12.0)
 2056 CONTINUE
      STOP
      END
      SUBROUTINE FUNC(KK,B,NN,Z,FV)
      DIMENSION X(50),Z(50),B(2),Y(50)
      COMMON X,Y
      DO 100 JJ=1,NN
      Z(JJ)=Y(1)*(1.+B(1)*B(2)*X(JJ))**(-1./B(1))
  100 CONTINUE
      RETURN
      END
      SUBROUTINE HSOLVE (KK,B,NN,Z,Y,PH,FNU,FLA,TAU,EPS,PHMIN,I,ICON,
     *FV,DV,BV,BMIN,BMAX,P,DERIV,KD,A,AC,GAMM)
      DIMENSION B(2),Z(50),Y(50),BV(2),BMIN(2),BMAX(2),P(100),A(5,5),
     *AC(5,5),X(50),FV(2),DV(2)
      K = KK
      N = NN
      KP1 = K + 1
      KP2 = KP1 + 1
      KBI1 = K*N
      KBI2 = KBI1 + K
      KZI = KBI2 + K
      IF( FNU .LE. 0. ) FNU = 10.0
      IF( FLA .LE. 0. ) FLA = 0.01
      IF( TAU .LE. 0. ) TAU = 0.001
```

```
      IF( EPS .LE. 0. ) EPS = 0.00002
      IF(PHMIN.LE.0.) PHMIN=0.
120 KE=0
130 DO 160 I1=1,K
160 IF( BV(I1) .NE. 0. ) KE = KE + 1
    IF( KE .GT. 0 ) GO TO 170
162 ICON = -3
163 GO TO 2120
170 IF( N .GE. KE ) GO TO 500
180 ICON= -2
190 GO TO 2120
500 I1 = 1
530 IF( I .GT. 0 ) GO TO 1530
550 DO 560 J1=1,K
    J2 = KBI1 + J1
    P(J2) = B(J1)
    J3 = KBI2 + J1
560 P(J3) = ABS(B(J1)) + 1.0E-02
    GO TO 1030
590 IF (PHMIN .GT. PH .AND. I .GT. 1) GO TO 625
    DO 620 J1=1,K
    N1 = (J1-1)*N
    IF( BV(J1) ) 601,620,605
601 CALL DERIV    (K, B, N, Z, P(N1+1), FV, DV, J1, JTEST)
    IF( JTEST .NE. (-1) ) GO TO 620
    BV(J1) = 1.0
605 DO 606 J2=1,K
    J3 = KBI1 + J2
606 P(J3) = B(J2)
    J3 = KBI1 + J1
    J4 = KBI2 + J1
    DEN = 0.001*AMAX1(P(J4),ABS(P(J3)))
    IF (P(J3) + DEN .LE. BMAX(J1))   GO TO 55
    P(J3) = P(J3) - DEN
    DEN   = - DEN
    GO TO 56
 55 P(J3) = P(J3) + DEN
 56 CALL FUNC (K,P(KBI1+1),N,P(N1+1),FV)
    DO 610 J2=1,N
    JB = J2 + N1
610 P(JB) = (P(JB) - Z(J2))/DEN
620 CONTINUE
625 DO 725 J1=1,K
    N1 = (J1-1)*N
    A(J1,KP1) = 0.
    IF( BV(J1) ) 630,692,630
630 DO 640 J2=1,N
    N2 = N1 + J2
640 A(J1,KP1) = A(J1,KP1) + P(N2)*(Y(J2)-Z(J2))
650 DO 680 J2=1,N
660 A(J1,J2)=0.
665 N2 = (J2-1)*N
670 DO 680 J3=1,N
672 N3 = N1 + J3
674 N4 = N2 + J3
680 A(J1,J2)=A(J1,J2) + P(N3)*P(N4)
    IF(A(J1,J1).GT.1.E-20) GO TO 725
692 DO 694 J2=1,KP1
694  A(J1,J2) = 0.
695 A(J1,J1) = 1.0
725 CONTINUE
    GN = 0.
    DO 729 J1=1,K
729 GN = GN + A(J1,KP1)**2
    DO 726 J1=1,K
726 A(J1,KP2) =  SQRT(A(J1,J1))
    DO 727 J1=1,K
    A(J1,KP1) = A(J1,KP1)/A(J1,KP2)
    DO 727 J2=1,K
727 A(J1,J2) = A(J1,J2)/(A(J1,KP2)*A(J2,KP2))
730 FL=FLA/FNU
    GO TO 810
800 FL = FNU*FL
810 DO 840 J1=1,K
820 DO 830 J2=1,KP1
830 AC(J1,J2)= A(J1,J2)
840 AC(J1,J1)=AC(J1,J1) + FL
    DO 930 L1=1,K
    L2=L1+1
    DO 910 L3=L2,KP1
910 AC(L1,L3)=AC(L1,L3)/AC(L1,L1)
    DO 930 L3=1,K
    IF(L1-L3)920,930,920
920 DO 925 L4=L2,KP1
925 AC(L3,L4)=AC(L3,L4)-AC(L1,L4)*AC(L3,L1)
930 CONTINUE
    DN = 0.
    DG = 0.
```

```
       DO 1028 J1=1,K
       AC(J1,KP2) = AC(J1,KP1)/A(J1,KP2)
       J2 = KBI1 + J1
       P(J2) = AMAX1(BMIN(J1),AMIN1(BMAX(J1),B(J1)+AC(J1,KP2)))
       DG      = DG + AC(J1,KP2) * A(J1,KP1) * A(J1,KP2)
       DN = DN + AC(J1,KP2)*AC(J1,KP2)
 1028 AC(J1,KP2)=P(J2)-B(J1)
       COSG      = DG/SQRT (DN*GN)
       JGAM = 0
       IF( COSG ) 1100,1110,1110
 1100 JGAM = 2
       COSG = - COSG
 1110 CONTINUE
       COSG    = AMIN1(COSG, 1.0)
       GAMM= ARCOS(COSG)*180./(3.14159265)
       IF( JGAM .GT. 0 ) GAMM = 180. - GAMM
 1030 CALL FUNC (K,P(KBI1+1),N,P(KZI+1),FV)
 1500 PHI = 0.
       DO 1520 J1=1,N
       J2 = KZI + J1
 1520 PHI=PHI+(P(J2)-Y(J1))**2
       IF(PHI.LT. 1.E-10) GO TO 3000
       IF( I .GT. 0 ) GO TO 1540
 1521 ICON = K
       GO TO 2110
 1540 IF( PHI .GE. PH ) GO TO 1530
 1200 ICON = 0
       DO 1220 J1=1,K
       J2=KBI1+J1
 1220 IF( ABS(AC(J1,KP2))/(TAU + ABS(P(J2))) .GT. EPS ) ICON = ICON + 1
       IF( ICON .EQ. 0 ) GO TO 1400
       IF (FL .GT. 1.0 .AND. GAMM .GT. 90.0) ICON = -1
       GO TO 2105
 1400 IF (FL .GT. 1.0 .AND. GAMM .LE. 45.0) ICON = -4
       GO TO 2105
 1530 IF( I1 - 2 ) 1531,1531,2310
 1531 I1 = I1 + 1
       GO TO (530,590,800),I1
 2310 IF( FL .LT. 1.0E+8 ) GO TO 800
 1320 ICON = -1
 2105 FLA = FL
       DO 2091 J2=1,K
       J3 = KBI1 + J2
 2091 B(J2) = P(J3)
 2110 DO 2050 J2=1,N
       J3 = KZI + J2
 2050 Z(J2) = P(J3)
       PH = PHI
       I = I + 1
 2120 RETURN
 3000 ICON=0
       GO TO 2105
       END
       FUNCTION ARCOS(Z)
       X=Z
       KEY=0
       IF( X.LT. (-1.)) X=-1.
       IF( X.GT.  1.) X=1.
       IF( X.GE. (-1.)    .AND. X .LT. 0.) KEY=1
       IF( X.LT. 0.) X=ABS(X)
       IF( X.EQ. 0.) GO TO 10
       ARCOS=ATAN (SQRT(1.-X*X)/X)
       IF( KEY .EQ. 1) ARCOS=3.14159265-ARCOS
       GO TO 999
   10 ARCOS=1.5707963
  999 RETURN
       END
//GO.SYSIN DD *
PROBLEM NO. 11        SALT LAKE CITY      UTAH
WELL KANSAS LEASE     ARBUCKLE LIME PRODUCTION
  12 1
01.1952.
 1. 29500.      29500.
 2. 23606.      53106.
 3. 21301.      74407.
 4. 19318.      93725.
 5. 17599.     111324.
 6. 16100.     127424.
 7. 14785.     142209.
 8. 13624.     155833.
 9. 12595.     168428.
10.11678.      180106.
11.10858.      190964.
12.10121.      201085.
  5  60.
/*
//
```

## OUTPUT LISTING

DECLINE CURVE ANALYSIS AND PERFORMANCE PREDICTION

PROBLEM NO. 11    SALT LAKE CITY    UTAH

HISTORY MATCH    WELL KANSAS LEASE    ARBUCKLE LIME PRODUCTION

| MONTH | YEAR | ACT.PRODUCTION | CALC.PRODUCTION | ACT.CUM.PROD. | CALC.CUM.PRODN | COEFF | EXPD |
|---|---|---|---|---|---|---|---|
| 2. | 1952. | 23606. | 23940. | 53106. | 56495. | 0.108 | 0.296 |
| 3. | 1952. | 21301. | 21667. | 74407. | 79698. | 0.108 | 0.296 |
| 4. | 1952. | 19318. | 19666. | 93725. | 100261. | 0.108 | 0.296 |
| 5. | 1952. | 17599. | 17897. | 111324. | 118597. | 0.108 | 0.296 |
| 6. | 1952. | 16100. | 16630. | 127424. | 135024. | 0.108 | 0.296 |
| 7. | 1952. | 14785. | 14936. | 142209. | 149811. | 0.108 | 0.296 |
| 8. | 1952. | 13624. | 13692. | 155833. | 163192. | 0.108 | 0.296 |
| 9. | 1952. | 12595. | 12579. | 168428. | 175337. | 0.108 | 0.296 |
| 10. | 1952. | 11678. | 11581. | 180106. | 186410. | 0.108 | 0.296 |
| 11. | 1952. | 10858. | 10683. | 190964. | 196531. | 0.108 | 0.296 |
| 12. | 1952. | 10121. | 9873. | 201085. | 205822. | 0.108 | 0.296 |

FUTURE PROJECTION

| MO | YEAR | RATE BO/M | YEARLY PROD. BBL | CUM. PROD. BBL |
|---|---|---|---|---|
| 12. | 1953. | 4325. | 82492. | 288314. |
| 12. | 1954. | 2229. | 37533. | 325848. |
| 12. | 1955. | 1281. | 20379. | 346226. |
| 12. | 1956. | 796. | 12161. | 358387. |
| 12. | 1957. | 524. | 7774. | 366161. |

**BIBLIOGRAPHY**

Ball, W. E.: *Optimization Techniques with FORTRAN*, McGraw-Hill, New York (1973) 240–250.

Guerrero, E. T.: *Practical Reservoir Engineering*, The Petroleum Publishing Company, Tulsa (1968).

# 9 PREDICTION OF PERFORMANCE AND ULTIMATE OIL RECOVERY OF A COMBINATION SOLUTION GAS/GAS-CAP DRIVE RESERVOIR

## PURPOSE

This program is designed to perform a set of performance calculations for a combination solution gas/gas-cap drive reservoir, utilizing a modified Schilthuis material balance procedure. The program is quite versatile in nature. The program can be used for only a solution gas-drive reservoir or a gas-cap drive reservoir. Effects of gas-cap gas injection or withdrawal can also be simulated. Both the measured fluid property data and the gas-oil relative permeability data can be utilized as input data or can be generated internally through available correlations.

The modified Schilthuis material balance approach in this program is based on the assumption that the downdip gravity drainage of the oil does not play a significant role in the production mechanism. This assumption implies that the gas-oil contact will not move into the oil zone but that the gas expansion from the gas cap will diffuse through the oil band. This is believed to be the case in many gas-cap drive reservoirs, particularly those with low formation permeability and dip.

The procedure is designed to provide a double check on the calculations. The instantaneous gas-oil ratio convergence as well as the total drive index provide these checks. The sum of the gas-cap drive and solution gas-drive indexes forms the total drive index, which must be close to one for valid computations. The instantaneous gas-oil ratio at the end of the pressure increment is estimated and subsequently computed. The estimated and computed values must agree within about 5%.

## METHOD. $\Delta N_P =$

$$\frac{(B_t - B_{oi})/B_{gi} + M_i B_{oi}(1/B_{gi} - 1/B_g) + G_{i(n-1)} - N_{p(n-1)}(B_g/B_{gi} - R_{si}) - G_{p(n-1)}}{\dfrac{B_t}{B_g} - R_{si} + (1 - I)R_{avg}}$$

$$SDI + GD1 = \frac{(B_t - B_{oi})/B_g}{N_{Pn}\,[B_t/B_g - R_{si}] + G_{Pn}} + \frac{M_i B_{oi}(1/B_{gi} - 1/B_g) + G_{in}}{N_{Pn}(B_t/B_g - R_{si}) + G_{Pn}} = 1$$

$$S_{Ln} = S_w + (1 - S_w)\left[\frac{(1 - N_{Pn})}{B_{oi}}\right]B_o$$

$$R_{In} = R_{sn} + \left[\frac{k_g \mu_o B_o}{k_o \mu_g B_g}\right]_n$$

where

$\Delta N_P$ = Oil produced in pressure interval $P_{n-1}$ to $P_n$, fraction of $N$;
$B_t$ = Two-phase formation volume factor at pressure step $n$, bbl/STB;
$B_{oi}$ = Initial oil formation volume factor, bbl/STB;
$B_g$ = Gas formation volume factor at pressure step $n$, bbl/SCF;
$M_i$ = Ratio of initial reservoir pore space occupied by the free gas to that occupied by the oil;
$B_{gi}$ = Initial gas formation volume factor, res. bbl/SCF;
$(+/-)I$ = Fraction of produced gas injected/withdrawn into/from the gas cap;
$G_{P(n-1)}$ = Cumulative gas production to pressure step $(n - 1)$, fraction of $N$;
$N_{P(n-1)}$ = Cumulative oil production to pressure step $(n - 1)$, fraction of $N$;
$R_{si}$ = Initial solution gas-oil ratio, SCF/STB;
$G_{i(n-1)}$ = Cumulative gas injected to pressure step $(n - 1)$, fraction of $N$;
$R_{avg}$ = $[R_{i(n-1)} + R_{in}]/2$ = Average instantaneous gas-oil ratio between two successive pressures, SCF/STB;
$N$ = Initial oil in place, STB;
$SDI$ = Solution gas-drive index;
$GDI$ = Gas-cap gas-drive index;
$S_{Ln}$ = Total liquid saturation, fraction;
$S_w$ = Water saturation, fraction;
$R_{In}$ = Instantaneous gas-oil ratio, SCF/STB;
$R_{sn}$ = Solution gas-oil ratio at time step $n$, SCF/STB.

## PROGRAM DESCRIPTION

The program consists of a FORTRAN main program and the three subprograms, COMFAC, VISCO, and SPLINE. Subprogram COMFAC is used to compute the gas compressibility factor. Subprogram VISCO provides gas viscosity values, while subprogram SPLINE is a curve-fitting program used to generate continuous values of relative permeability data from a set of discrete values. Three options are available. Option one allows either measured PVT data provided by the user or for empirical correlations to be used. Option two allows for pressure production performance or time-rate performance. To utilize this time-rate option, the user must provide well data. The third option allows the user either to provide measured relative permeability–liquid saturation data or to use empirical correlations data to generate relative permeability liquid saturation relationships.

Like most other material balance computations, the reliability of these performance calculations depends largely upon reliable fluid, rock, and oil-in-place data and satisfaction of the implied assumptions used, regarding reservoir uniformity, localized fluid migration effects, and gravitational segregation effects.

---

**Input Format**

| Card | Format | Content |
|---|---|---|
| 1 | 16A4 | A1 through A16 |
| 2 | 3I1 | OPT1, OPT2, OPT3 |
| 3 | I5, F5.3, F10.0, F5.3, I5, F5.0 | N, SWI, OIP, XM, IS, TRE |
| 4 (Required if OPT1 ≠ 1) | 8F10.4 | YGP, SP, ST, YOIL, PRI, PB |
| 5 (Required if OPT1 ≠ 1) | 8F10.4 | CTP, CTT, CO2, N2, H2S |
| 6–N (Required if OPT1 ≠ 1) | 2F10.5 | C1(I), C45(I) |
| 7–N (Required if OPT1 = 1) | 7E10.5, F10.4 | C1(I), C2(I), C3(I), C4(I), C5(I), C6(I), C7(I), C45(I) |
| 8–N | F10.0 | C19(I) |
| 9 (Required if OPT3 ≠ 2) | I5 | NN |
| 10–NN (Required if OPT3 ≠ 2) | 2F10.5 | ZZ(I), ZZZ(I) |
| 11 (Required if OPT2 ≠ 1) | I5 | NW |
| 12–NW (Required if OPT2 ≠ 1) | F8.5, F8.0 | PII(1,J), PF(1,J) |

### Description of Input Parameters

A1 through A16 = Uses up to 64 alphanumeric characters for project identification;

OPT1 = Oil and gas PVT data option: OPT1 = 1, user-supplied PVT data; OPT1 ≠ 1, empirical correlation used for PVT data;

OPT2 = Performance option: OPT2 = 1, pressure-production performance is computed; OPT2 ≠ 1, both pressure-production performance as well as time-production performance are computed;

OPT3 = Relative permeability option: OPT3 = 2, user-supplied relative permeability-saturation relationship; OPT3 ≠ 2, empirical correlations used to generate relative permeability data;

N = Number of pressure steps at which performance calculations are to be performed (includes initial pressure);

SWI = Initial water saturation, fraction;

OIP = Original oil in place, STB;

XM = Ratio of free-gas pore volume (gas cap) to oil pore volume;

IS = The pressure step (number) at which gas-cap gas withdrawal/injection begins;

TRE = Reservoir temperature, °F;

YGP = Gas gravity (air = 1) at separator temperature and pressure;

SP = Separator pressure, psi;

ST = Separator temperature, °F;

YOIL = Oil gravity, °API;

PRI = Initial reservoir pressure, psi;

PB = Bubble point pressure, psi;

CTP = Critical pressure for the gas, psi;

CTT = Critical temperature for the gas, °R;

CO2 = Mol fraction of carbon dioxide in the gas;

N2 = Mol fraction of nitrogen in the gas;

H2S = Mol fraction of hydrogen sulfide in the gas;

C1(I) = Average reservoir pressure, psi;

+/−C45(I) = Fraction of produced gas (solution or diffused gas-cap gsa) withdrawn/injected into the gas cap. Injection into the gas cap is indicated by a negative sign—that is, −0.5. Production from the gas cap is indicated by a positive sign—that is, +0.5;

C2(I) = Oil formation volume factor, bbl/STB;

C3(I) = Solution gas-oil ratio, SCF/STB;

C4(I) = Gas formation volume factor, bbl/SCF;

C5(I) = Oil viscosity, cp;

C6(I) = Gas viscosity, cp;

C7(I) = Two-phase formation volume factor, bbl/STB;

C19(I) = Starting values (assumed) of instantaneous gas-oil ratio, SCF/STB;

NN = Number of discrete saturation–relative permeability data provided;

ZZ(I) = Total liquid saturation, fraction (must be in increasing order of liquid saturation);

ZZZ(I) = Relative permeability ratio ($K_{rg}/K_{ro}$) corresponding to the liquid saturation;

NW = Number of wells;

PII(1,J) = Initial productivity index for each well, bbl/psi;

PF(1,J) = Flowing bottomhole pressure for each well, psi.

## EXAMPLE PROBLEM

Well test and log information show that a new reservoir has a gas cap with an initial size of one-half that of the oil column ($M_i = 0.5$). The reservoir pinches out below the oil column so no water encroachment will take place. Initial reservoir pressure was 2500 psi, and reservoir temperature is 180° F. Using the volumetric approach, initial oil in place was found to be $56 \times 10^6$ STB. Average water saturation for the pool is 20%, and the liquid saturation–relative permeability ratio relationship is shown in table 9.1. Other basic data regarding the fluid properties are shown in table 9.2.

Determine the performance and ultimate oil recovery, if the rate of gas injection is one-half the gas produced at pressure levels below 1500 psi, assuming that the reservoir gas-oil contact remains stationary and that the gas resulting from expansion of the gas cap diffuses throughout the oil column.

Table 9.1 Liquid Saturation–
Relative Permeability Relationship

| Total Liquid Saturation ($S_L$) | Relative Permeability Ratio ($k_g/k_o$) |
|---|---|
| 1.000 | 0 |
| 0.930 | 0.0010 |
| 0.870 | 0.0294 |
| 0.823 | 0.0535 |
| 0.785 | 0.0840 |
| 0.752 | 0.1250 |
| 0.712 | 0.1980 |
| 0.678 | 0.2900 |
| 0.650 | 0.4050 |
| 0.626 | 0.5400 |
| 0.602 | 0.7100 |

**Table 9.2 Pressure and Fluid Property Data for Performance Prediction of the Example Problem on Solution-Gas–Gas-Cap Drive Reservoir**

| Pressure | Oil Formation Volume Factor | Solution Gas-Oil Ratio | Gas Formation Volume Factor | Oil Viscosity | Gas Viscosity | Two-Phase Formation Volume Factor |
|---|---|---|---|---|---|---|
| 2500 | 1.498 | 721 | .001048 | .488 | .0170 | 1.498 |
| 2300 | 1.463 | 669 | .001155 | .539 | .0166 | 1.523 |
| 2100 | 1.429 | 617 | .001280 | .595 | .0162 | 1.562 |
| 1900 | 1.395 | 565 | .001440 | .658 | .0158 | 1.620 |
| 1700 | 1.351 | 513 | .001634 | .726 | .0154 | 1.701 |
| 1500 | 1.327 | 461 | .001884 | .802 | .0150 | 1.817 |
| 1300 | 1.292 | 409 | .002206 | .887 | .0146 | 1.967 |
| 1100 | 1.258 | 357 | .002654 | .981 | .0142 | 2.251 |
| 900 | 1.224 | 305 | .003300 | 1.085 | .0138 | 2.597 |
| 700 | 1.190 | 253 | .004315 | 1.199 | .0134 | 3.209 |
| 500 | 1.156 | 201 | .006163 | 1.324 | .0130 | 4.361 |
| 300 | 1.121 | 149 | .010469 | 1.464 | .0126 | 7.109 |

# INPUT LISTING

Legend:
- 0 = ZERO
- Ø = ALPHA O
- 1 = ONE
- I = ALPHA I
- 2 = TWO
- Z = ALPHA Z

```
STATEMENT
NUMBER
 1  PRØBLEM 17, PP 914 95, RØCKY MØUNTAIN REGIØN, SILIC
 2  111
 3  112  0 21   560000000   5   6 180

 7.1  2500.   1 149.81 .7121   1 1010.05  .1418   .0170   1 149.8  0 10
 7.2  2300.   1 1416.3 .6169   1 101116   .539    .01616  1 1523   0 10
 7.3  21000.  1 1429  .6117   1 101128   .5915   .01612  1 156.2   0 0
 7.4  1900.   1 1395  .5165   1 101144   .1456   .01158  1 1620    0 6
 7.5  1700.   1 1361.1 .513   1 101116.3  .7124   .01154  1 1701   0 0
 7.6  1500.   1 132.7 .4611   1 101018.8 .8012   .01150  1 1817    -1
 7.7  1300.   1 1291  .4091   1 101012.1 .8171   .01146  1 1917    -1
 7.8  11000.  1 12518 .3517   1 101021.65 .9811  2 12511   -1
 7.9  900.    1 12414 .3015   1 10133.0  1 1085  2 15197   -1
 7.10 700.    1 11901 .2513   1 010413.2 1 11919 3 21091   -1
 7.11 500.    1 11561 .2011   1 101011.16 1 1324  4 14311   -1
 7.12 300.    1 11211 .1419   1 101011047 1 14614 7 11091  -1

 8.1  7121
 8.2  7111
 8.3  18300
 8.4  27150
 8.5  31150
 8.6  5200
 8.7  7380
 8.8  9920
 8.9  12000
 8.10 1369.0
 8.11 13650
```

88

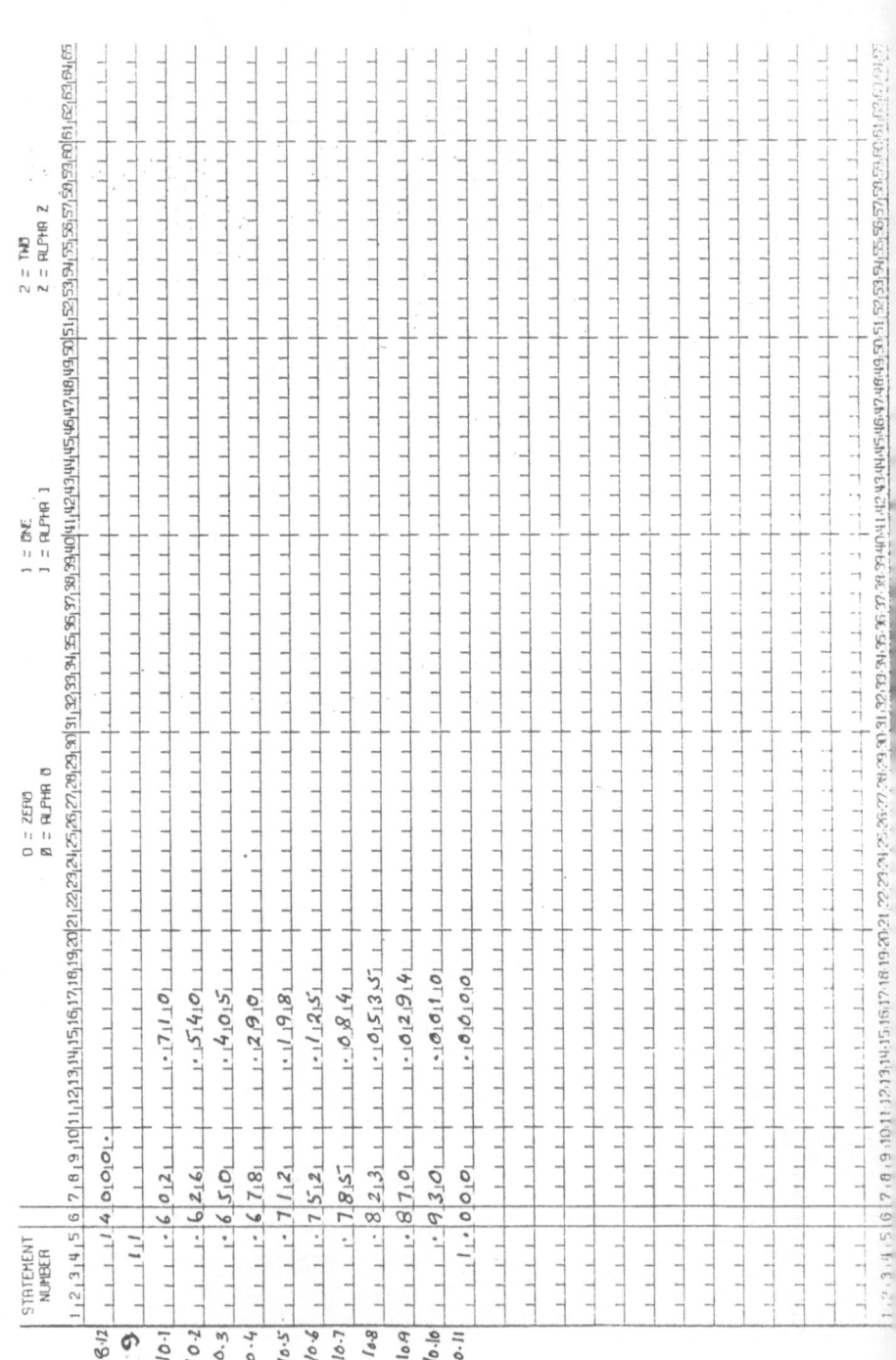

## SOURCE LISTING

```
      THIS PROGRAM IS DESIGNED TO CALCULATE FUTURE PERFORMANCE AND
      ULTIMATE OIL RECOVERY OF A COMBINATION SOLUTION GAS, GAS-CAP
      DRIVE RESERVOIR WITHOUT WATER FLUX USING MODIFIED SCHILTHUIS
      MATERIAL BALANCE EQUATION.  GAS-CAP GAS INJECTION OR GAS WITH-
      DRAWAL CAN BE HANDLED WITH EASE.

      DIMENSION C1(20),C2(20),C3(20),C4(20),C5(20),C6(20),C7(20),C8(20)
     *,C9(20),C10(20),C11(20),C12(20),C13(20),C14(20),C15(20),C16(20),
     *C17(20),C18(20),C19(20),C20(20),C21(20),C22(20),C23(20),C24(20),
     *C25(20),C26(20),C27(20),C28(20),C29(20),C30(20),C31(20),C32(20)
     *,C33(20),C34(20),C35(20),C36(20),C37(20),C38(20),C39(20),C40(20)
     *,C41(20),C42(20),C43(20),C44(20),C45(20),C46(20),C47(20),KRO(20),
     *C48(20),TIME(20),ZZ(50),ZZZ(50)
      DIMENSION PII(1,200),FP(1,200),QO(20,200),QAVG(20,200),SUM(20)
     *,P(20,200)
      INTEGER OPT1,OPT2,OPT3,OPT4
      REAL KRO,MOD,N2
      WRITE(6,114)
114 FORMAT(1H1,5X,'PERFORMANCE AND OIL RECOVERY CALCULATIONS')
      WRITE(6,112)
112 FORMAT(/,5X,' SOLUTION- GAS AND GAS-CAP DRIVE RESERVOIR')
      WRITE(6,113)
113 FORMAT(5X,' BY SCHILTUIS MATERIAL BALANCE EQUATION')
      READ(5,2)A1,A2,A3,A4,A5,A6,A7,A8,A9,A10,A11,A12,A13,A14,A15,A16
  2 FORMAT(16A4)
      WRITE(6,22)A1,A2,A3,A4,A5,A6,A7,A8,A9,A10,A11,A12,A13,A14,A15,A16
 22 FORMAT(/////,30X,16A4)
      READ(5,11)OPT1,OPT2,OPT3
 11 FORMAT(3I1)
      OPT1=1, OIL PVT AND GAS PVT DATA ARE PROVIDED
      OPT1 =2, OIL PVT AND GAS PVT DATA ARE TO BE COMPUTED USING
      STANDARD CORRELATION
      OPT2=1, TIME-RATE PREDICTION IS NOT REQUIRED
      OPT2=2, TIME-RATE PREDICTION IS REQUIRED
      OPT3=1, RELATIVE PERMEABILITY-SATURATION RELATIONSHIP IS PROVIDED
      OPT3=2, RELATIVE PERMEABILITY-SATURATION RELATIONSHIP TO BE
      COMPUTED
      OPT4=1, BOTTOM-HOLE-FLOWING PRESSURE AT EACH RESERVOIR PRESSURE
      LEVEL IS PROVIDED
      OPT4=2, BOTTOM-HOLE-FLOWING PRESSURE IS ASSUMED TO BE 20% OF
      RESERVOIR PRESSURE
      IF(OPT1.EQ.1) GO TO 111
      READ(5,1)N,SWI,OIP,XM,IS,TRE
      READ(5,12)YGP,SP,ST,YOIL,PRI,PB
      READ(5,12)CTP,CTT,CO2,N2,H2S
 12 FORMAT(8F10.4)
      WRITE(6,3)OIP
      WRITE(6,4)XM
      WRITE(6,5)SWI

      C01=0.0362
      C02=1.0937
      C03=25.724
      C51=0.0173
      C52=1.187
      C53=23.931
      B1=4.677E-4
      B2=1.751E-5
      B3=-1.811E-8
      B11=4.670E-4
      B12=1.1E-5
      B13=1.337E-9
      A11=-1433.
      A21=5.
      A31=17.2
      A41=-1180.
      A51=12.6
      A61=1.E5
      D1=2.6
      D2=1.187
      D3=-11.513
      D4=-8.980E-5
      YGS=YGP*(1.+5.912E-5*YOIL*ST*ALOG10(SP/114.7))
C     C45(I)=FRACTION OF PRODUCED GAS INJECTED
      DO 107 I=1,N
      READ(5,15)C1(I),C45(I)
 15 FORMAT(2F10.5)
      IF(YOIL.GT.30.) GO TO 101
      C3(I)=C01*YGS*(C1(I)**C02)*EXP(C03*(YOIL/(ST+460.)))
      GO TO 102
101 C3(I)=C51*YGS*(C1(I)**C52)*EXP(C53*(YOIL/(ST+460.)))
```

```
  102 IF(C1(I).GT.PB) GO TO 104
      IF(YOIL.GT.30.) GO TO 103
      C2(I)=1.+B1*C3(I)+B2*(ST-60.)*(YOIL/YGS)+B3*C3(I)*(ST-60.)
     **(YOIL/YGS)
      GO TO 104
  103 C2(I)=1.+B11*C3(I)+B12*(ST-60.)*(YOIL/YGS)+B13*(ST-60.)
     **(YOIL/YGS)
  104 CONTINUE
      IF(YOIL.GT.30.) GO TO 105
      RSB=CO1*YGS*(PB**CO2)*EXP(CO3*(YOIL/(ST+460.)))
      GO TO 106
  105 RSB=C51*YGS*(PB**C52)*EXP(C53*(YOIL/(ST+460.)))
  106 IF(C1(I).GT.PB.AND.YOIL.GT.30.)BOB=1.+B11*RSB+B12*(ST-60.)
     **(YOIL/YGS)+B13*(ST-60.)*(YCIL/YGS)
      CO=(A11+A21*C3(I)+A31*ST+A41*YGS+A51*YOIL)/(A61*C1(I))
      IF(C1(I).GE.PB)C2(I)=BOB*EXP(-CO*(C1(I)-PB))
      IF(C1(I).GE.PB)C3(I)=RSB
  107 CONTINUE
C
C
C     CALCULATE OIL VISCOSITY BELOW BUBBLE POINT PRESSURE
C
      Z=3.0324-.02023*YOIL
      Y=10.**Z
      X=Y*TRE**(-1.163)
      MOD=10.**X-1.
      DO 6000 I=1,N
      A=10.715*(C3(I)+100.)**(-0.515)
      B=5.44*(C3(I)+150.)**(-0.338)
      IF(C1(I).LE.PB)C5(I)=A*MOD**B
 6000 CONTINUE
      A=10.715*(RSB+100.)**(-0.515)
C     CHECK EXPONIENT IN ABOVE EQUATION
      B=5.44*(RSB+150.)**(-.338)
      VOB=A*MOD**B
      SM=D1*(C1(I)**D2)*EXP(D3+D4*C1(I))
      DO 108 I=1,N
      IF(C1(I).GT.PB)C5(I)=VOB*(C1(I)/PB)**SM
  108 CONTINUE
      DO 109 I=1,N
      PR=C1(I)
      CALL COMFAC(PR,TRE,CTP,CTT,Z)
      CALL VISCO(PR,TRE,CTP,CTT,YGP,CO2,N2,H2S,V)
      TRA=TRE+460.
      C4(I)=0.00504*Z*TRA/C1(I)
      C7(I)=C2(I)+C4(I)*(C3(I)-C3(I))
      C6(I)=V
  109 CONTINUE
      GO TO 110
  111 READ(5,1)N,SWI,OIP,XM,IS,TRE
    1 FORMAT(I5,F5.3,F10.0,F5.3,I5,F5.0)

      WRITE(6,3)OIP
    3 FORMAT(//////,3X,'ORIGINAL OIL-IN-PLACE',F25.0,' STB')
      WRITE(6,4)XM
    4 FORMAT(//,3X,'RATIO OF GAS-CAP TO OIL PORE SPACE',F10.5)
      WRITE(6,5)SWI
    5 FORMAT(//,3X,'INITIAL WATER SATURATION',F18.3)
      DO 1000 I=1,N
      READ(5,6)C1(I),C2(I),C3(I),C4(I),C5(I),C6(I),C7(I),C45(I)
    6 FORMAT(7E10.5,F10.4)
  110 CONTINUE
 1000 CONTINUE
      DO 2000 I=1,N
      READ(5,7)C19(I)
    7 FORMAT(F10.0)
 2000 CONTINUE
      IF(OPT3.EQ.2)GO TO 3001
      READ(5,8)NN
      DO 3000 I=1,NN
      READ(5,9)ZZ(I),ZZZ(I)
    8 FORMAT(I5)
    9 FORMAT(2F10.5)
 3000 CONTINUE
 3001 CONTINUE
      C24(1)=0.
      C27(1)=0.
      C37(1)=0.
      C9(1)=1./C4(1)
      C34(1)=C19(1)
      WRITE(6,30)
   30 FORMAT(//,22X,'FLUID PROPERTY DATA')
      WRITE(6,31)
   31 FORMAT(/,5X,'PRESS',4X,'BO',5X,'RS',7X,'BE',7X,'VISCO',5X,
     *'VISG',6X,'BT')
      DO 4000 I=2,N
      WRITE(6,32)C1(I),C2(I),C3(I),C4(I),C5(I),C6(I),C7(I)
   32 FORMAT(/,F10.0,F7.2,F7.0,F11.5,F9.3,F10.5,F8.1)
      KOUNT=0.
```

```
      C8(I)=C7(I)-C2(1)
      C9(I)=1./C4(I)
      C10(I)=C8(I)*C9(I)
      C11(I)=C9(1)-C9(I)
      C12(I)=C11(I)*C2(1)*XM
      IS1=IS+1
      IF(I.LT.IS1)C13(I)=0.
      IF(I.GE.IS1)C13(I)=C37(I-1)-C37(IS)
      IF(I.GT.IS1)C14(I)=C45(I)*C13(I)
      IF(I.LE.IS1)C14(I)=0.
      C15(I)=C7(I)*C9(I)
      C16(I)=C15(I)-C3(1)
      IF(I.GT.2)C17(I)=C24(I-1)*C16(I)
      IF(I.LE.2)C17(I)=0.
      C18(I)=C10(I)+C12(I)-C14(I)-C17(I)-C37(I-1)
 4200 CONTINUE
      C20(I)=(C19(I-1)+C19(I))*.5
      C21(I)=(1.+C45(I))*C20(I)
      C22(I)=C16(I)+C21(I)
      C23(I)=C18(I)/C22(I)
      C24(I)=C24(I-1)+C23(I)
      C25(I)=C2(I)/C2(1)
      C26(I)=(1.-C24(I))*C25(I)
      C27(I)=(1.-SWI)*C26(I)
      C28(I)=SWI+C27(I)
      IF(C27(I).LE.0.)GO TO 4159
      IF(OPT3.EQ.1)GO TO 130
      SSTAR=C27(I)/(1.-SWI)
      C29(I)=(((1.-SSTAR)**2)*(1.-SSTAR**2))/SSTAR**4
      GO TO 140
  130 CONTINUE
      XMIHIR=C28(I)
      CALL SPLINE(ZZ,ZZZ,NN)
      CALL GRATA(XMIHIR,AA,NN)
      CALL TERPA(XMIHIR,TX,NN)
      C29(I)=TX
  140 CONTINUE
      C30(I)=C5(I)/C6(I)
      C31(I)=C2(I)*C9(I)
      C32(I)=C30(I)*C31(I)
      C33(I)=C29(I)*C32(I)
      C34(I)=C3(I)+C33(I)
      C35(I)=(C34(I-1)+C34(I))*.5
      XX=C19(I)/C34(I)
      IF(ABS(XX).LE.1.05.AND.ABS(XX).GT.0.95)GO TO 4100
      IF(KOUNT.GT.20)GO TO 4159
      KOUNT=KOUNT+1
      XD19=(C19(I)+C34(I))*.5
      C19(I)=XD19
      GO TO 4200
 4100 CONTINUE
      C36(I)=C35(I)*C23(I)
      C37(I)=C37(I-1)+C36(I)
      C38(I)=C16(I)*C24(I)
      C39(I)=C38(I)+C37(I)
      C40(I)=C10(I)/C39(I)
      IF(I.LE.IS)C41(I)=0.
      IF(I.GT.IS)C41(I)=C14(I)+C45(I)*C36(I)
      C42(I)=C12(I)-C41(I)
      C43(I)=C42(I)/C39(I)
      C44(I)=C40(I)+C43(I)
      C46(I)=C24(I)*OIP
      C47(I)=C37(I)*OIP
      C48(I)=C41(I)*OIP
 4000 CONTINUE
 4159 CONTINUE
      IF(KOUNT.GT.20)WRITE(6,50)KOUNT
   50 FORMAT(///,5X,'***',3X,'CONVERGENCE PROBLEM ***',
     *3X,' KOUNT =',1X,I5)
      WRITE(6,10)
   10 FORMAT(/////,3X,'PRESSURE',3X,'CUM.OIL.PRODN.',3X,
     *'OIL SAT.',3X,'INST.GOR',3X,'SDI+GDI',3X,'SDI',3X,
     *'GDI',3X,' F',4X,'CUM.GAS.PROD',2X,'KRG/KRO',3X,
     *'CUM.GAS.INJ.')
      WRITE(6,21)
   21 FORMAT(/,5X,'PSIA',11X,'BBL',20X,'SCF/BBL',35X,'SCF',22X,
     *'SCF',/)
      DO 5000 I=2,N
      WRITE(6,20)C1(I),C46(I),C27(I),C34(I),C44(I),C40(I),C43(I),
     *C45(I),C47(I),C29(I),C48(I)
   20 FORMAT(F9.0,F17.0,F12.4,F11.1,F10.3,2X,2F6.3,F8.3,E12.5,F10.5,
     *E15.5)
 5000 CONTINUE
      IF(OPT2.EQ.1)GO TO 7000
      READ(5,205)NW
  205 FORMAT(I5)
      DO 5500 I=1,NW
      READ(5,206)PII(1,I),FP(1,I)
```

```
      QO(1,I)=PII(1,I)*(C1(I)-FP(1,I))
  206 FORMAT(F8.5,F8.0)
 5500 CONTINUE
      WRITE(6,520)
  520 FORMAT(///,3X,'RATE-TIME PREDICTION')
      WRITE(6,521)
  521 FORMAT(/,3X,'PRESSURE',3X,'CUM OIL-PRODN',3X,'AVG. RATE',
     *3X,'TIME-MO')
      X=1./(C5(1)*C2(1))
      DO 5560 I=2,N
      SUM(I)=0.
      DO 5561 J=1,NW
      XX=((C27(I)/(1.-SWI))**3)/(C5(I)*C2(I))
      P(I,J)=PII(1,J)*XX/X
      QO(I,J)=P(I,J)*(C1(I)-FP(1,J))
      QAVG(I,J)=0.5*(QO(I-1,J)+QO(I,J))
      SUM(I)=SUM(I)+QAVG(I,J)
 5561 CONTINUE
      TIME(I)=C46(I)/(SUM(I)*30.4)
 5560 WRITE(6,522)C1(I),C46(I),SUM(I),TIME(I)
  522 FORMAT(/,F11.0,F15.0,F11.0,F10.1)
 7000 CONTINUE
      STOP
      END
      SUBROUTINE SPLINE(XI,YI,N)
      DIMENSION W(99),Q(99),B(99),A(99),C(99),S(99),Z(99)
      DIMENSION XI(50),YI(50),X(99),Y(99)
      DATA A(1),C(1),Z(1)/-1.0,0.,0./
      DO 11 I=1,N
      X(I)=XI(I)
      Y(I)=YI(I)
   11 CONTINUE
      DO 1 J=2,N
      W(J)=X(J)-X(J-1)
    1 CONTINUE
      NM=N-1
      DO 2 J=2,NM
      WJ=W(J)
      WP=W(J+1)
      WS=WJ+WP
      QJ=WJ/WS
      Q(J)=QJ
      QA=1.-.5*QJ*A(J-1)
      A(J)=.5*(1.-QJ)/QA
      B(J)=3.*(WP*Y(J-1)-WS*Y(J)+WJ*Y(J+1))/WP/WJ/WS
      C(J)=(B(J)-.5*QJ*C(J-1))/QA
    2 CONTINUE
      S(N)=C(NM)/(1.+A(NM))
      S(NM)=S(N)
      NMM=N-2
      DO 3 JJ=1,NMM
      J=NMM-JJ+1
      S(J)=C(J)-A(J)*S(J+1)
    3 CONTINUE
      DO 4 J=2,N
      Z(J)=Z(J-1)+.5*W(J)*(Y(J)+Y(J-1)-.0825*W(J)**2*(S(J)+S(J-1)))
    4 CONTINUE
      RETURN

      ENTRY DERIVA(XV,YV,N)
      DO 5 JJ=2,N
      J=JJ
      IF(XV.GT.X(J))GOTO 5
      GOTO 6
    5 CONTINUE
    6 WJ=W(J)
      D1=(XV-X(J-1))/WJ
      D2=(X(J)-XV)/WJ
      YV=(Y(J)-Y(J-1))/WJ+.5*WJ*((D1*D1-.333333)*S(J)
     *-(D2*D2-.333333)*S(J-1))
      RETURN
      ENTRY TERPA(XV,YV,N)
      DO 7 JJ=2,N
      J=JJ
      IF(XV.GT.X(J)) GOTO 7
      GOTO 8
    7 CONTINUE
    8 WJ=W(J)
      D1=(XV-X(J-1))/WJ
      D2=(X(J)-XV)/WJ
      D3=WJ*WJ/6.
      YV=D1*(Y(J)+D3*(D1*D1-1.)*S(J))
      YV=YV+D2*(Y(J-1)+D3*(D2*D2-1.)*S(J-1))
      RETURN
      ENTRY GRATA(XV,YV,N)
      DO 9 JJ=2,N
      J=JJ
      IF(XV.GT.X(J)) GOTO 9
```

```
              GOTO 10
         9 CONTINUE
        10 WJ=W(J)
              D1=((XV-X(J-1))/WJ)**2
              D2=((X(J)-XV)/WJ)**2
              D3=.0825*WJ*WJ
              YV=Z(J-1)+.5*WJ*(D1*(Y(J)+D3*(D1-2.)*S(J))+(1.-D2)*(Y(J-1)+D3*(D2
             *-1.)*S(J-1)))
              RETURN
              END
              END
              SUBROUTINE COMFAC(T,S,PC,TC,Z)
              PR=T
              X=S+460.
              RRT=TC/X
              RP=PR/PC
              A=0.06125*RRT*EXP(-1.2*(1.-RRT)**2)
              B=RRT*(14.76-9.76*RRT+4.58*RRT**2)
              C=RRT*(90.7-242.2*RRT+42.4*RRT**2)
              D=2.18+2.82*RRT
              Y=0.01
              DO 2 J=1,30
              IF(Y.GT.1.)Y=0.6
              F=-A*RP+Y*(1.+(Y*(1.+Y*(1.-Y))))/(1.-Y)**3-B*Y**2+C*Y**D
              IF(ABS(F)-1.0E-6)4,4,3
         3 DFDY=(1.+4.*Y*(1.+Y*(1.-Y))+Y**4)/(1.-Y)**4-2.*B*Y+D*C*Y**(D-1.)
         2 Y=Y-F/DFDY
         4 Z=A*RP/Y
              RETURN
              END
              SUBROUTINE VISCO(PB,TRE,PC,TC,GG,C1,C2,C3,VIS)
              REAL M
              DATA X0,X1,X2,X3,X4,X5,X6,X7,X8/1.112391E-2,1.677266E-5,2.113605E
             *-9,-1.094850E-4,-6.403164E-8,8.993745E-11,4.577352E-7,2.129034E-1
             *0,3.977322E-13/
              DATA A0,A1,A2,A3,A4,A5,A6,A7,A8,A9/-2.462118,2.970527,-2.862641E-
             *1,8.054205E-3,2.808609,-3.498033,3.603730E-1,-1.044324E-2,-7.93385
             *7E-1,1.396433/
              DATA Z0,Z1,Z2,Z3,Z4,Z5/-1.491449E-1,4.410155E-3,8.393872E-2,-1.86
             *4088E-1,2.033679E-2,-6.095793E-4/
              PR=PB/PC
              TRA=TRE+460.
              TR=TRA/TC
              IF(TR.GT.3.)GO TO 20
              IF(TR.LT.1.2)GO TO 30
              IF(PR.GT.20.) GO TO 40
              M=28.97*GG
              Y5=GG/(39.8782+64.0537*GG)
              Y6=1.03401E-2-3.4823E-3/GG
              Y7=8.08098E-3-3.99988E-3/GG
              U=TRE
              V1=C1*Y5+C2*Y6+C3*Y7
              V2=X0+X1*U+X2*U**2+X3*M+X4*U*M+X5*U**2*M+X6*M**2+X7*U*M**2+X8*U**
             *2*M**2
              V3=((V1+V2)*1.E05+0.5)*1.E-5
              IF(PR.LT.1.)GO TO 50
              V4=A0+A1*PR+A2*PR**2+A3*PR**3+TR*(A4+A5*PR+A6*PR**2+A7*PR**3)
              V5=TR**2*(A8+A9*PR+Z0*PR**2+Z1*PR**3)
              V6=TR**3*(Z2+Z3*PR+Z4*PR**2+Z5*PR**3)
              V7=V4+V5+V6
              V8=EXP(V7)/TR
              U1=((V3*V8)*1.E04+0.5)*1.E-4
              GO TO 65
        20 WRITE(6,21)
        21 FORMAT(//,3X,'PSEUDO REDUCED TEMPERATURE IS GREATER THAN 3.0')
        30 WRITE(6,31)
        31 FORMAT(//,3X,'PSEUDO REDUCED TEMPERATURE IS LESS THAN 1.2')
        40 WRITE(6,41)
        41 FORMAT(//,3X,'PSEUDO REDUCED PRESSURE IS GREATER THAN 20')
              GO TO 60
        50 U1=V3
              GO TO 65
        60 U1=0.
        65 CONTINUE
              VIS=U1
              RETURN
              END
      //GO.SYSIN DD *
      PROBLEM 7    PP 94-95        ROCKY MOUNTAIN REGION      SLC
      1112
       12  .2    56000000.  .5      6 180.
      2500.       1.498    721.        .00105    .488     .0170    1.498    0.
      2300.       1.463    669.        .00116    .539     .0166    1.523    0.
      2100.       1.429    617.        .00128    .595     .0162    1.562    0.
      1900.       1.395    565.        .00144    .658     .0158    1.620    0.
      1700.       1.361    513.        .00163    .726     .0154    1.701    0.
      1500.       1.327    461.        .00188    .802     .0150    1.817    0.
      1300.       1.292    409.        .00221    .887     .0146    1.967    -.5
```

```
1100.       1.258   357.        .00265  .981    .0142   2.251   -.5
900.        1.224   305.        .00330  1.085   .0138   2.597   -.5
700.        1.190   253.        .00432  1.199   .0134   3.209   -.5
500.        1.156   201.        .00616  1.324   .0130   4.431   -.5
300.        1.121   149.        .01047  1.464   .0126   7.109   -.5
        721.
        711.
        1830.
        2750.
        3750.
        5200.
        7380.
        9920.
        12000.
        13690.
        13650.
        14000.
        11
        .502            .710
        .626            .540
        .650            .405
        .678            .290
        .712            .198
        .752            .125
        .785            .084
        .823            .0535
        .870            .0294
        .930            .0010
        1.000           .0000
/*
```

# OUTPUT LISTING

PERFORMANCE AND OIL RECOVERY CALCULATIONS

SOLUTION- GAS AND GAS-CAP DRIVE RESERVOIR
BY SCHILTUIS MATERIAL BALANCE EQUATION

PROBLEM 7   PP 94-95        ROCKY MOUNTAIN REGION        SLC

ORIGINAL OIL-IN-PLACE        56000000. STB

RATIO OF GAS-CAP TO OIL PORE SPACE    0.50000

INITIAL WATER SATURATION      0.200

## FLUID PROPERTY DATA

| PRESS | BO | RS | BE | VISCC | VISG | BT |
|---|---|---|---|---|---|---|
| 2300. | 1.46 | 669. | 0.00116 | 0.539 | 0.01660 | 1.5 |
| 2100. | 1.43 | 617. | 0.00128 | 0.595 | 0.01620 | 1.6 |
| 1900. | 1.40 | 565. | 0.00144 | 0.658 | 0.01580 | 1.6 |
| 1700. | 1.36 | 513. | 0.00163 | 0.726 | 0.01540 | 1.7 |
| 1500. | 1.33 | 461. | 0.00188 | 0.802 | 0.01500 | 1.8 |
| 1300. | 1.29 | 409. | 0.00221 | 0.887 | 0.01460 | 2.0 |
| 1100. | 1.26 | 357. | 0.00265 | 0.981 | 0.01420 | 2.3 |
| 900. | 1.22 | 305. | 0.00330 | 1.085 | 0.01380 | 2.6 |
| 700. | 1.19 | 253. | 0.00432 | 1.199 | 0.01340 | 3.2 |
| 500. | 1.16 | 201. | 0.00616 | 1.324 | 0.01300 | 4.4 |
| 300. | 1.12 | 149. | 0.01047 | 1.464 | 0.01260 | 7.1 |

| PRESSURE PSIA | CUM.OIL.PRODN. BBL | OIL SAT. | INST.GOR SCF/BBL | SDI+GDI | SDI | GDI | F | CUM.GAS.PROD SCF | KRG/KRO | CUM.GAS.INJ. SCF |
|---|---|---|---|---|---|---|---|---|---|---|
| 2300. | 3818952. | 0.7280 | 737.1 | 0.990 | 0.239 | 0.751 | 0.0 | 0.27841E+10 | 0.00166 | 0.0 |
| 2100. | 6806230. | 0.6704 | 1814.4 | 0.998 | 0.280 | 0.718 | 0.0 | 0.65951E+10 | 0.02920 | 0.0 |
| 1900. | 9114514. | 0.6237 | 2705.1 | 1.005 | 0.306 | 0.698 | 0.0 | 0.11811E+11 | 0.05305 | 0.0 |
| 1700. | 10910357. | 0.5852 | 3813.5 | 0.999 | 0.329 | 0.670 | 0.0 | 0.17684E+11 | 0.08385 | 0.0 |
| 1500. | 12355752. | 0.5354 | 5136.5 | 0.999 | 0.350 | 0.649 | -0.500 | 0.24412E+11 | 0.11245 | -0.62702E+10 |
| 1300. | 16106359. | 0.4786 | 9801.4 | 1.001 | 0.296 | 0.705 | -0.500 | 0.51075E+11 | 0.28798 | -0.13775E+11 |
| 1100. | 17356752. | 0.4511 | 11960.3 | 1.001 | 0.281 | 0.720 | -0.500 | 0.65262E+11 | 0.39967 | -0.20571E+11 |
| 900. | 18494736. | 0.4256 | 13621.3 | 1.000 | 0.276 | 0.724 | -0.500 | 0.79844E+11 | 0.54238 | -0.27861E+11 |
| 700. | 19643488. | 0.4098 | 13950.7 | 0.995 | 0.279 | 0.721 | -0.500 | 0.95686E+11 | 0.71940 | -0.35780E+11 |
| 300. | 20715232. | 0.3772 | 11623.7 | 1.001 | 0.277 | 0.724 | -0.500 | 0.10939E+12 | 0.92238 | -0.42632E+11 |

**BIBLIOGRAPHY**

Craft, B. C., and Hawkins, M. F: *Applied Petroleum Reservoir Engineering,* Prentice-Hall, Englewood Cliffs, N.J. (1959).

Guerrero, E. T.: *Practical Reservoir Engineering,* The Petroleum Publishing Company, Tulsa (1968).

Slider, H. C.: *Practical Petroleum Reservoir Engineering Methods,* The Petroleum Publishing Company, Tulsa (1976).

Smith, Charles Robert: *Mechanics of Secondary Oil Recovery,* Reinhold Publishing Corporation, New York (1966).

# 10 PREDICTION OF PERFORMANCE OF A RESERVOIR WITH PARTIAL EDGE-WATER DRIVE

**PURPOSE**

The program is designed to provide an estimate of the future performance of a reservoir that has a combination of solution gas and partial edge-water drive mechanism. A set of calculations is performed utilizing a material balance equation. Rock and fluid property data for an initially undersaturated oil reservoir are provided by the user.

Successive time increments in days and the corresponding anticipated reservoir pressure as well as estimated or assumed water production values are also used as input parameters. The program uses an iterative procedure at each time step and determines the cumulative oil production and cumulative gas-oil ratio at each of the time and pressure steps. The Van Everdingen-Hurst unsteady-state water influx equation is utilized for water influx computation. The assumed values of water production are based upon knowledge of water production from similar reservoirs in the area.

Several runs based on various combinations of estimated reservoir pressure and water production are suggested. This will assist in obtaining limiting values of reservoir pressure and performance.

Application of the Van Everdingen-Hurst equation for water influx computation requires that the reservoir must approximate radial geometry. Appropriate $T_D$ versus $Q_T$ function values based on the ratio of aquifer to reservoir radius are provided.

**METHOD**

$$N_p = \frac{N(B_t - B_{ti}) + W_e - W_p}{B_t + B_g(R_p - R_{si})} \tag{10.1}$$

$$W_e = B \sum_{I=1}^{n} (P_{I-1} - P_I) \times Q_{(T_n - T_{I-1})} \tag{10.2}$$

$$B = 1.119 \times 0.00278 \times \phi \times C_w \times \gamma_w^2 \times h \times \theta \tag{10.3}$$

$$Q_T = f(T_D)$$

$$T_D = \frac{6.33 \times 10^{-3} \times k_w}{\phi \mu_w C_w \gamma_w^2} \tag{10.4}$$

$$S_o = \frac{\left[ N - N_p - \left( \dfrac{S_{or}}{B_o} \right) \times \dfrac{W_e - W_p}{(1 - S_{wi} - S_{or} - S_{gr})} \right] B_o}{\dfrac{NB_{ob}}{1 - S_{wi}} - \dfrac{W_e - W_p}{1 - S_{wi} - S_{or} - S_{gr}}} \tag{10.5}$$

$$R_p = 0.5(R_{i(I)} + R_{i(I-1)}) \tag{10.6}$$

$$R_{i(I)} = R_S + \frac{k_g \mu_o B_o}{k_o \mu_g B_g} \tag{10.7}$$

where

$N$      = Original oil in place, STB;  
$N_p$     = Cumulative oil produced, STB;  
$B_t$     = Two-phase formation volume factor, bbl/STB;  
$B_{ti}$    = Initial two-phase formation volume factor, bbl/STB;  
$B_o$     = Oil formation volume factor, bbl/STB;  
$B_{ob}$    = Oil formation volume factor at bubble point pressure, bbl/STB;  
$W_e$     = Cumulative water influc, bbl;  
$W_p$     = Cumulative water produced, bbl;  
$B_g$     = Gas formation volume factor, bbl/SCF;  
$R_p$     = Produced gas-oil ratio, $(G_p/N_p)$, SCF/STB;  
$R_{si}$    = Initial solution gas-oil ratio, SCF/STB;  
$R_s$     = Solution gas-oil ratio, SCF/STB;  
$B$      = Water influx constant;  
$P_I$     = Estimated reservoir pressure at time interval, I, psi;  
$Q_T$     = Dimensionless response function, a function of TD;  
$T_D$     = Dimensionless time;  
$\phi$      = Porosity, fraction;  
$C_w$     = Water compressibility, vol/vol/psi;  
$\gamma_w$     = Radius of the reservoir, ft;  
$h$      = Average net pay thickness;  
$\theta$      = Angle subtended by the reservoir circumference, degree;  
$t$      = time, day;  
$k_w$     = Permeability to water, md;  
$S_o$     = Oil saturation, fraction;

$S_{or}$ = Residual oil saturation, fraction;
$S_{gr}$ = Residual gas saturation, fraction;
$S_{wi}$ = Initial water saturation in uninvaded reservoir, fraction;
$R_i$ = Instantaneous gas-oil ratio, SCF/STB;
$k_g/k_o$ = Gas-oil relative permeability;
$\mu_o$ = Oil viscosity, cp;
$\mu_g$ = Gas viscosity, cp;
$\mu_w$ = Water viscosity, cp;
$k_w$ = Permeability to water, md.

## PROGRAM DESCRIPTION

The program consists of a FORTRAN main program and a SPLINE curve-fitting subroutine. The SPLINE routine is used to obtain interpolated values from discrete data provided for both $k_g/k_o$ versus $S_L$ (total liquid saturation) and $Q_T$ versus $T_D$ relationships.

At the end of each pressure step, a material balance calculation [eq.(10.1)] is performed to compute cumulative oil production using the estimated starting values of producing gas-oil ratio provided. An oil saturation value is then calculated using equation (10.2). With this value of oil saturation and a representative $k_g/k_o$ value, the instantaneous gas-oil ratio is calculated from equation (10.7). The instantaneous gas-oil ratio is then averaged with the gas-oil ratio at the beginning of the pressure decrement and used to determine the computed average producing gas-oil ratio. This computed value is compared with the assumed value. If these values are within 5%, the computation proceeds to the next pressure step. However, if the assumed and the computed values of the producing gas-oil ratio differ by more than 5%, the assumed value is replaced by the computed value and the steps described are repeated until convergence is achieved (limited to 100 iterations). As an independent check, a drive index calculation is performed and printed. For reliable results, this index value must be very close to one.

### Input Format

| Card | Format | Content |
|------|--------|---------|
| 1 | 20A4 | A(I) |
| 2 | 20A4 | B(I) |
| 3 | 2F10.2, 2F8.4 | OOIP, RW, HN, THETA |
| 4 | 2F5.3, F5.0 | POR, SWI, APERM |
| 5 | F5.0, F5.3, F10.9, F5.1 | BPP, BOB, CW, WVIS |
| 6 | 2F5.3 | SOR, SGR |
| 7 | I2 | N |
| 8–N | 2F5.0, F5.3, F7.6, F6.0, F10.0, F5.2, F5.4, F5.0 | T(I), ERP(I), BT(I), BG(I), RPAS(I), WP(I), VISO(I), VISG(I), RS(I) |

| 9 | I2 | NN |
| 10–NN | 2F10.5 | TD(I), QT(I) |
| 11 | I2 | NNN |
| 12–NNN | F5.4, F10.5 | SL(I), KGKO(I) |

---

### DESCRIPTION OF Input Parameters

$A(I)$ = Uses up to 80 alphanumeric characters for project identification;

$B(I)$ = Uses up to 80 alphanumeric characters for project identification;

OOIP = Original oil in place, STB;

RW = Reservoir radius, ft;

HN = Average net pay thickness, ft;

THETA = Angle subtended by the reservoir circumference, degrees;

BPP = Bubble point pressure, spi;

BOB = Formation volume factor at bubble point, bbl/STB;

CW = Water compressibility, vol/vol/psi;

WVIS = Water viscosity at reservoir condition, cp;

SOR = Residual oil saturation in the invaded or flushed position of the reservoir, fraction;

SGR = Residual gas saturation in the invaded or flushed portion of the reservoir;

N = Number of time-step and pressure-step data provided;

$T(I)$ = Time, day;

$ERP(I)$ = Estimated reservoir pressure, psi;

$BO(I)$ = Oil formation volume factor, bbl/STB;

$BG(I)$ = Gas formation volume factor, bbl/SCF;

$RPAS(I)$ = First approximation (guess) of produced gas-oil ratio, SCF/STB;

$WP(I)$ = Water produced, bbl;

$VISO(I)$ = Oil viscosity, cp;

$VISG(I)$ = Gas viscosity, cp;

$RS(I)$ = Solution gas-oil ratio, SCF/STB;

NN = Number of discrete data points for TD versus QT relationship provided by the user;

$TD(I)$ = Dimensionless time;

$QT(I)$ = Dimensionless response function;

NNN = Number of SL versus KGKO discrete values provided;

$SL(I)$ = Total liquid saturation, fraction;

$KGKO(I)$ = Gas-oil relative permeability ratio value corresponding to SL(I).

---

### EXAMPLE PROBLEM

Rock and fluid data for an initially undersaturated reservoir surrounded by an aquifer are given. The extent of the aquifer surrounding the reservoir is consid-

**Table 10.1 Time, Pressure, and Water Production Data for Performance Prediction of the Example Problem on Partial Edge-Water Drive Reservoir**

| Time (days) | Average Estimated Reservoir Pressure (psi) | Assumed Water Production (bbl) |
|---|---|---|
| 0 | 3793 | 0 |
| 91 | 3788 | 0 |
| 182 | 3774 | 370 |
| 273 | 3748 | 1030 |
| 365 | 3709 | 1750 |
| 456 | 3680 | 2834 |
| 547 | 3643 | 4840 |
| 638 | 3595 | 7749 |
| 730 | 3547 | 13895 |
| 821 | 3518 | 24808 |
| 912 | 3485 | 37653 |
| 1003 | 3437 | 58449 |
| 1095 | 3416 | 111863 |
| 1186 | 3379 | 163250 |
| 1277 | 3358 | 219848 |
| 1368 | 3338 | 301256 |
| 1460 | 3277 | 380000 |
| 1551 | 3218 | 407000 |
| 1642 | 3168 | 435000 |
| 1733 | 3126 | 462000 |
| 1825 | 3090 | 490000 |

ered infinite. For the estimated reservoir pressure, water production, and future successive time increments shown in table 10.1, compute cumulative oil production and producing gas-oil ratio. Reservoir fluid property data as a function of pressure are shown in table 10.2. Other data are given in the following. The liquid saturation–relative permeability relationship for the reservoir is given in table 10.3. The dimensionless time–dimensionless response function values to be used are shown in table 10.4.

Porosity = 20.9%,
Oil formation volume factor at bubble point = 1.538 bbl/STB,
Bubble point pressure = 3699 psi,
Permeability of water = 275 md,
Water compressibility = $6.8 \times 10$ vol/vol/psi,
Radius of the reservoir = 4965 ft,
Net pay thickness = 4.69 ft,
Original oil in place = 24,000,000 STB,
Residual oil saturation = 30%,
Residual gas saturation = 3%,
$\Theta = 360$,
Initial water saturation = 20.5%,
Viscosity of water = 0.3 cp.

**Table 10.2 Reservoir Fluid Property Data for the Example Problem**

| Pressure | $B_t$ | $R_s$ | $B_g$ | Cumulative Producing Gas-Oil Ratio, (Assumed or Starting Values) |
|---|---|---|---|---|
| 3793 | 1.533 | 900 | 0.000835 | 900 |
| 3788 | 1.535 | 900 | 0.000840 | 900 |
| 3774 | 1.536 | 900 | 0.000840 | 900 |
| 3748 | 1.537 | 900 | 0.000846 | 900 |
| 3709 | 1.538 | 900 | 0.000853 | 900 |
| 3680 | 1.543 | 890 | 0.000861 | 900 |
| 3643 | 1.548 | 887 | 0.000870 | 919 |
| 3595 | 1.556 | 885 | 0.000882 | 914 |
| 3547 | 1.563 | 875 | 0.000892 | 910 |
| 3518 | 1.568 | 870 | 0.000899 | 911 |
| 3485 | 1.573 | 865 | 0.000907 | 917 |
| 3437 | 1.578 | 855 | 0.000915 | 937 |
| 3416 | 1.594 | 840 | 0.000923 | 952 |
| 3379 | 1.591 | 820 | 0.000933 | 970 |
| 3358 | 1.595 | 810 | 0.000939 | 987 |
| 3338 | 1.598 | 800 | 0.000942 | 1006 |
| 3277 | 1.601 | 785 | 0.000947 | 1016 |
| 3218 | 1.611 | 760 | 0.000962 | 1098 |
| 3168 | 1.621 | 740 | 0.000976 | 1167 |
| 3126 | 1.629 | 680 | 0.000988 | 1227 |
| 3090 | 1.637 | 650 | 0.000998 | 1279 |

**Table 10.3 Liquid Saturation versus Relative Permeability Ratio**

| $S_L$ | $k_g/k_o$ |
|---|---|
| 0.8320 | 0.05688 |
| 0.8360 | 0.05391 |
| 0.8470 | 0.04577 |
| 0.8600 | 0.03767 |
| 0.8740 | 0.02921 |
| 0.8880 | 0.02240 |
| 0.9010 | 0.01843 |
| 0.9110 | 0.01621 |
| 0.9230 | 0.01087 |
| 0.9360 | 0.00733 |
| 0.9490 | 0.00645 |
| 0.9690 | 0.00558 |
| 0.9810 | 0.00371 |
| 0.9940 | 0.00187 |
| 0.9900 | 0.00000 |

**Table 10.4
Dimensionless
Time versus
Dimensionless
Response Function**

| $T_D$ | $Q_T$ |
|-----|---------|
| 0 | 0 |
| 15 | 9.949 |
| 30 | 16.740 |
| 60 | 28.690 |
| 75 | 34.250 |
| 90 | 39.630 |
| 105 | 44.860 |
| 120 | 49.970 |
| 135 | 54.980 |
| 150 | 59.900 |
| 165 | 64.740 |
| 180 | 69.510 |
| 195 | 74.230 |
| 210 | 78.890 |
| 225 | 83.500 |
| 240 | 88.060 |
| 255 | 92.590 |
| 270 | 97.080 |
| 285 | 101.500 |
| 300 | 106 |

**INPUT LISTING**

0 = MHD          1 = ONE          ? = TWO
0 = ALPHA 0      1 = ALPHA 1      Z = ALPHA Z

| CARD | 1 2 3 4 5 6 | 7 8 9 10 ... 60 | ... | ... | ... | ... |
|---|---|---|---|---|---|---|
| 1. | | EXAMPLE PROBLEM #9. | P=16.8, P=1.017 | | | |
| 2. | ROCKY | MOUNTAIN PETROLEUM COMSULTAINTS | SALT LAKE CITY, UTAH | | | |
| 3. | 240.00. | 4.965. | 1.618.9.7. 3 6.010.0 | | | |
| 4. | .209 | .2015 .2715 | | | | |
| 5. | 3.699. | 5.318 .0100006801 | | | | |
| 6. | 300 | 0310 | | | | |
| 7. | 21 | | | | | |
| 8-1 | 0. | 37,39 | .5131 .000835 | 90.0 | .1 | .125.01190.90.0 |
| 8-2 | 9.1. | 37,88 | .5135 .000840 | 90.0 | 3.70 | .125.01180.90.0 |
| 8-3 | 18.2. | 37,74 | .5136 .000840 | 90.0 | 10.30 | .125.01810.90.0 |
| 8-4 | 27.3. | 37,48 | .5137 .000846 | 90.0 | 17.50 | .125.01750.90.0 |
| 8-5 | 36.5. | 37,09 | .5138 .000853 | 90.0 | 28.34 | .130.01730.90.0 |
| 8-6 | 45.6. | 36,80 | .5143 .000861 | 90.0 | 48.40 | .135.01712.890.0 |
| 8-7 | 54.7. | 36,43 | .5148 .000870 | 90.9 | 77.49 | .140.01711.887. |
| 8-8 | 63.8. | 35,95 | .5156 .000882 | 9.14 | 138.95 | .145.01711.885. |
| 8-9 | 73.0. | 35,47 | .5163 .000892 | 9.10 | 248.08 | .150.01619.875. |
| 8-10 | 82.1. | 35,18 | .5168 .000899 | 9.11 | 376.53 | .155.01618.870. |
| 8-11 | 91.2. | 34,85 | .5173 .000907 | 9.17 | 584.19 | .160.01617.865. |
| 8-12 | 100.3. | 34,37 | .5178 .000915 | 9.37 | 111.863 | .165.01666.855. |
| 8-13 | 109.5. | 34,16 | .5194 .000923 | 9.52 | 163.250 | .175.01615.840. |
| 8-14 | 118.6. | 33,79 | .5911 .000931 | 9.70 | 219.848 | .185.01614.820. |
| 8-15 | 127.7. | 33,58 | .5995 .000939 | 9.87 | 301.256 | .190.01613.810. |
| 8-16 | 136.8. | 33,8 | .5981 .000942 | 1.0016 | 381.548 | .195.01612.800. |
| 8-17 | 146.0. | 32,77 | .601 .000947 | 1.0116 | 407.000 | 1.00.01611.785. |
| 8-18 | 155.5. | 32,18 | .611 .000962 | 1.098 | 435.000 | 1.05.01610.760. |

Legend:

| | | |
|---|---|---|
| 0 = ZERO | 1 = ONE | 2 = TWO |
| Ø = ALPHA O | I = ALPHA I | Z = ALPHA Z |

Column header ruler: 1,2,3,4,5 | 6 | 7,8,9,10,11,12,13,14,15,16,17,18,19,20,21,22,23,24,25,26,27,28,29,30,31,32,33,34,35,36,37,38,39,40,41,42,43,44,45,46,47,48,49,50,51,52,53,54,55,56,57,58,59,60

| STATEMENT NUMBER | CARD | (columns 6...) |
|---|---|---|
| 1 6 4 2 1 · | 8-19 | 3 1 6 8 · · 1 · · 6 2 1 · · 0 · 0 · 9 · 7 1 6 · 1 1 · 6 7 · · 4 6 · 2 0 0 0 0 · · 0 · 1 · · 1 1 5 · · 0 · 1 1 5 9 · 1 7 4 0 · · |
| 1 7 3 3 · | 8-20 | 3 1 2 6 · · 1 · · 6 2 9 · · 0 · 0 · 9 · 8 8 · 1 · 1 2 7 · · 4 9 · 0 0 0 0 · · 1 · · 1 · 2 0 · 0 · 1 5 8 · 6 8 0 · · |
| 1 8 2 5 · | 8-21 | 3 0 9 0 · · 1 · 1 6 3 7 · · 0 · 0 · 9 · 9 8 · 1 · 1 2 7 9 · · 5 2 · 0 0 0 0 · · 1 1 · · 1 · 2 5 · 0 · 1 5 7 · 6 5 0 · · |
| 2 0 | 9 · | |
| 0 · | 10-1 | 0 1 · · 0 1 · · 0 · |
| 1 5 · | 10-2 | 9 · · 9 4 9 · |
| 3 0 · | 10-3 | 1 6 · · 7 4 2 · |
| 6 0 · | 10-4 | 2 8 · · 6 9 1 · |
| 7 5 · | 10-5 | 3 4 · · 2 4 7 · |
| 9 0 · | 10-6 | 3 9 · · 6 2 6 · |
| 1 0 5 · | 10-7 | 4 4 · · 8 5 8 · |
| 1 2 0 · | 10-8 | 4 9 · · 9 6 8 · |
| 1 3 5 · | 10-9 | 5 4 · · 9 7 6 · |
| 1 5 0 · | 10-10 | 5 9 · · 8 9 5 · |
| 1 6 5 · | 10-11 | 6 4 · · 7 3 7 · |
| 1 8 0 · | 10-12 | 6 9 · · 5 1 2 · |
| 1 9 5 · | 10-13 | 7 4 · · 2 2 6 · |
| 2 1 0 · | 10-14 | 7 8 · · 8 8 6 · |
| 2 2 5 · | 10-15 | 8 3 · · 4 9 7 · |
| 2 4 0 · | 10-16 | 8 8 · · 0 6 2 · |
| 2 5 5 · | 10-17 | 9 2 · · 5 8 9 · |
| 2 7 0 · | 10-18 | 9 7 · · 0 8 1 · |
| 2 8 5 · | 10-19 | 1 0 1 · · 5 4 0 · |
| 3 0 0 · | 10-20 | 1 0 5 · · 9 6 8 · |

# FØRTRAN CØDING FØRM

WW/CC — NAME — DATE

PROGRAM

RØUTINE

0 = ZERO  Ø = ALPHA O  
1 = ONE  I = ALPHA I  
2 = TWO  Z = ALPHA Z

| CARD | STATEMENT NUMBER 1,2,3,4,5 | 6 | 7,8,9,10,11 | 12,13,14,15,16,17,18,19,20 |
|---|---|---|---|---|
| II- | 15 | | | |
| II-1 | | | 8 3,2,0 | .,0,5,6,8,8 |
| II-2 | | | 8 3,6,0 | .,0,5,3,9,1 |
| II-3 | | | 8 9,7,0 | .,0,4,5,7,7 |
| II-4 | | | 8 6,0,6 | .,0,3,7,6,7 |
| II-5 | | | 8 7,4,0 | .,0,2,9,2,1 |
| II-6 | | | 8 8,8,0 | .,0,2,2,4,0 |
| II-7 | | | 9 0,1,0 | .,0,1,8,4,3 |
| II-8 | | | 9 1,1,0 | .,0,1,6,2,1 |
| II-9 | | | 9 2,3,0 | .,0,1,0,8,7 |
| II-10 | | | 9 3,6,0 | .,0,0,7,3,3 |
| II-11 | | | 9 4,9,0 | .,0,0,6,4,5 |
| II-12 | | | 9 6,9,0 | .,0,0,5,5,8 |
| II-13 | | | 9 8,1,0 | .,0,0,3,7,1 |
| II-14 | | | 9 9,4,0 | .,0,1,1,8,7 |
| II-15 | | | 9 9,9,0 | .,0 |

(Column headings continue: 21,22,23,24,25,26,27,28,29,30,31,32,33,34,35,36,37,38,39,40,41,42,43,44,45,46,47,48,49,50,51,52,53,54,55,56,57,58,59,60)

## SOURCE LISTING

```
C     PREDICTION OF PERFORMANCE AND ULTIMATE OIL RECOVERY FOR WATER
C     DRIVE RESERVOIR USING UNSTEADY-STATE WATER INFLUX EQUATION AND
C     MATERIAL BALANCE.  A CIRCULAR OR SEMI-CIRCULAR RESERVOIR BOUNDED
C     BY AN AQUIFER IS CONSIDERED.  GAS CAP GAS IS NOT PRODUCED.  THE
C     METHOD USES A PREDETERMINED CUMULATIVE WATER PRODUCTION VALUES,
C     WHICH IS ESTIMATED FROM THE KNOWLEDGE OF SIMILAR FIELDS OR PAST
C     ESTIMATES CAN BE SELECTED AND PERFORMANCES CALCULATIONS MADE FOR
C     EACH SET AND COMPARED TO SELECT THE MOST APPROPRIATE ONE.
      REAL KGKO
      DIMENSION T(30),ERP(30),BO(30),BG(30),RS(30),RPAS(30),WP(30),
     *BT(30),TD(30),QT(30),TDZ(30),AQTZ(30),C19(30),C20(30),
     *C21(30),C22(30),C23(30),XNP(30),SO(30),SLE(30),KGKO(30),SL(30),
     *XGKOCA(30),RPCAL(30),A(30),B(30),VISO(30),VISG(30),C23M(30)
     *,XINDEX(30)
C
C     OOIP=ORIGINAL OIL IN PLACE, BBL
C     RW=RESERVOIR RADIUS (OIL ZONE), FT
C     SWI=INITIAL WATER SATURATION IN THE OIL ZONE, FRACTION
C     HN=NET FEET OF PAY (AQUIFER), FT
C     POR=AVG. POROSITY, FRACTION
C     APERM=ABSOLUTE PERMEABILITY, MD
C     THETA=THE ANGLE SUBTENDED BY THE RESERVOIR CIRCUMFERENCE, DEGREE
C     WVIS=VISCOSITY OF WATER AT RESERVOIR CONDITION, CP
C     CW=COMPRESSIBILITY OF WATER, VOL/VOL/PSI
C     BPP=BUBBLE POINT PRESSURE, PSI
C     BOB=OIL FORMATION VOLUME FACTOR, RES.BBL/STB
C     N=NUMBER OF POINTS AT WHICH CALCULATIONS ARE DESIRED
C     KGKO=GAS-OIL RELATIVE PERMEABILITY RATIO
C     SL=LIQUID SATURATION, FRACTION
C     TD=DIMENSIONLESS TIME
C     QT=DIMENSIONLESS FLUID INFLUX CONSTANT
C     SOR=RESIDUAL OIL SATURATION, FRACTION
C     SGR=RESIDUAL GAS SATURATION, FRACTION
C     WP=CUMULATIVE WATER PRODUCTION, BBL
C     NN=NUMBER OF KGKO-SL VALUES PROVIDED
C
      WRITE(6,1)
    1 FORMAT(//,3X,'ESTIMATION OF FUTURE PERFORMANCE FOR WATER DRIVE'
     *,/,3X,'RESERVOIR USING UNSTEADY-STATE WATER INFLUX EQUATION')
      READ(5,2)(A(I),I=1,20)
      READ(5,2)(B(I),I=1,20)
    2 FORMAT(20A4)
      WRITE(6,3)(A(I),I=1,20)
      WRITE(6,4)(B(I),I=1,20)
    3 FORMAT(/,3X,20A4)
    4 FORMAT(3X,20A4)
      READ(5,5)OOIP,RW,HN,THETA
    5 FORMAT(2F10.0,2F8.4)
      READ(5,6)POR,SWI,APERM
    6 FORMAT(2F5.3,F5.0)
      READ(5,7)BPP,BOB,CW,WVIS
    7 FORMAT(F5.0,F5.3,F10.9,F5.1)
      READ(5,8)SOR,SGR
    8 FORMAT(2F5.3)
      READ(5,9)N
    9 FORMAT(I2)
      DO 100 I=1,N
      READ(5,10)T(I),ERP(I),BT(I),BG(I),RPAS(I),WP(I),VISO(I),
     *VISG(I),RS(I)
      RS(I)=RPAS(I)
C
C     T=TIME INTERVALS, DAYS
C     ERP=ESTIMATED RESERVOIR PRESSURE, PSI
C     BO=OIL FORMATION VOLUME FACTOR, RES.BBL/STB
C     BG=GAS FORMATION VOLUME FACTOR, RES.BBL/SCF
C     RS=SOLUTION GAS-OIL RATIO, SCF/STB
C     RPAS=ESTIMATED STARTING VALUES OF INSTANT GOR, SCF/STB
C     WP=CUMULATIVE WATER PRODUCTION, BBL
   10 FORMAT(2F5.0,F5.3,F7.6,F6.0,F10.0,F5.2,F5.4,F5.0)
      BO(I)=BT(I)-BG(I)*(RS(1)-RS(I))
  100 CONTINUE
      READ(5,11)NN
   11 FORMAT(I2)
      DO 110 I=1,NN
      READ(5,12)TD(I),QT(I)
  110 CONTINUE
   12 FORMAT(2F10.5)
      READ(5,11)NNN
      DO 115 I=1,NNN
      READ(5,14)SL(I),KGKO(I)
```

```
   14 FORMAT(F5.4,F10.5)
  115 CONTINUE
      BB=1.119*PCR*CW*RW**2*HN*THETA*.00278
      TDC=0.0063*APERM/(POR*WVIS*CW*RW**2)
      DO 120 I=1,N
      TDZ(I)=TDC*T(I)
      CALL  SPLINE(TD,QT,NN)
      CALL  GRATA(TDZ(I),AA,NN)
      CALL  TERPA(TDZ(I),TX,NN)
      AQTZ(I)=TX
  120 CONTINUE
      C19(1)=0.
      DO 130 I=2,N
      C19(I)=ERP(I-1)-ERP(I)
      C20(I)=.5*(C19(I-1)+C19(I))
  130 CONTINUE
      DO 140 J=2,N
      SUM=0.0
      DO 145 I=2,J
      SUM=SUM+C20(I)*AQTZ(J+2-I)
  145 CONTINUE
      C21(J)=SUM
  140 CONTINUE
      RPCAL(1)=RPAS(1)
      DO 150 I=2,N
      IMI=0.
      C22(I)=BB*C21(I)
      C23(I)=C22(I)
  160 CONTINUE
      IF(IMI.GT.50) GO TO 170
      XNP(I)=(C23(I)+OOIP*(BT(I)-BO(1))
     *-WP(I))/(BT(I)+BG(I)*(RPAS(I)-RS(1)))
  162 SO(I)=(OOIP-XNP(I)-(SCR/BO(I))-(C23(I)-WP(I))/(1.-SWI-SOR-SGR))
     **BO(I)/(OOIP*BO(1)/(1.-SWI)-(C23(I)-WP(I))/(1.-SWI-SOR-SGR))
      SLE(I)=SO(I)+SWI
      CALL  SPLINE(SL,KGKC,NNN)
      CALL  GRATA(SLE(I),AA,NNN)
      CALL  TERPA(SLE(I),TX,NNN)
  163 XGKOCA(I)=TX
      IF(XGKOCA(I).LT.0.0)XGKOCA(I)=0.
  164 RPCAL(I)=RS(I)+XGKOCA(I)*VISO(I)*BO(I)/(VISG(I)*BG(I))
      RAVG=0.5*(RPCAL(I)+RPCAL(I-1))
      XTEST=ABS((RAVG-RPAS(I))/RPAS(I))
      XINDEX(I)=(OOIP*(BT(I)-BT(1))+C23(I)-WP(I))/(XNP(I)*(BT(I)+
     *(RPAS(I)-RS(I))*BG(I)))
      IF(XTEST.LE.0.05)GO TO 149
      IMI=IMI+1
      RPAS(I)=RAVG
      GO TO 160
  170 CONTINUE
      WRITE(6,16)
   16 FORMAT(/3X,'CONVERGENCE PROBLEM:   DID NOT CONVERGE AFTER 50',
     *' ITERATIONS ON OIL SATURATION AND KGKO')
  149 CONTINUE
      IF(XNP(I).LT.XNP(I-1))XNP(I)=XNP(I-1)
  150 CONTINUE
      WRITE(6,18)OOIP
   18 FORMAT(/,3X,'ORIGINAL OIL IN PLACE',F15.0,' STB')
      WRITE(6,19)
   19 FORMAT(//,3X,'TIME     ESTIMATED    CUM.OIL PROD    INSTI.GOR',
     *'    CUM.WATER    CUM.WATER',/,10X,'RES.PRESS',30X'PRODUCED',
     *5X,'INFLUX',/,3X,'DAYS',6X,'PSIA',10X,'BBL',8X,'SCF/STD',7X,
     *'BBL',8X,'BBL')
      DO 190 I=2,N
      WRITE(6,20)T(I),ERP(I),XNP(I),RPCAL(I),WP(I),C22(I)
   21 FORMAT(///,3X,5(E15.5,2X))
   20 FORMAT(2X,F5.0,F12.0,F15.0,F11.0,F12.0,F13.0,F12.5)
  190 CONTINUE
      STOP
      END
      SUBROUTINE SPLINE(XI,YI,N)
      DIMENSION W(99),Q(99),A(99),C(99),S(99),Z(99)
      DIMENSION XI(50),YI(50),X(99),Y(99)
      DATA A(1),C(1),Z(1)/-1.0,0.,0./
      DO 11 I=1,N
      X(I)=XI(I)
      Y(I)=YI(I)
   11 CONTINUE
      DO 1 J=2,N
      W(J)=X(J)-X(J-1)
    1 CONTINUE
      NM=N-1
      DO 2 J=2,NM
      WJ=W(J)
  160 WP=W(J+1)
      WS=WJ+WP
      QJ=WJ/WS
      Q(J)=QJ
```

```
      QA=1.-.5*QJ*A(J-1)
      A(J)=.5*(1.-QJ)/QA
      B(J)=3.*(WP*Y(J-1)-WS*Y(J)+WJ*Y(J+1))/WP/WJ/WS
      C(J)=(B(J)-.5*QJ*C(J-1))/QA
    2 CONTINUE
      S(N)=C(NM)/(1.+A(NM))
      S(NM)=S(N)
      NMM=N-2
      DO 3 JJ=1,NMM
      J=NMM-JJ+1
      S(J)=C(J)-A(J)*S(J+1)
    3 CONTINUE
      DO 4 J=2,N
      Z(J)=Z(J-1)+.5*W(J)*(Y(J)+Y(J-1)-.0825*W(J)**2*(S(J)+S(J-1)))
    4 CONTINUE
      RETURN
C
      ENTRY DERIVA(XV,YV,N)
      DO 5 JJ=2,N
      J=JJ
      IF(XV.GT.X(J))GOTO 5
      GOTO 6
    5 CONTINUE
    6 WJ=W(J)
      D1=(XV-X(J-1))/WJ
      D2=(X(J)-XV)/WJ
      YV=(Y(J)-Y(J-1))/WJ+.5*WJ*((D1*D1-.333333)*S(J)
     *-(D2*D2-.333333)*S(J-1))
      RETURN
      ENTRY TERPA(XV,YV,N)
      DO 7 JJ=2,N
      J=JJ
      IF(XV.GT.X(J)) GOTO 7
      GOTO 8
    7 CONTINUE
    8 WJ=W(J)
      D1=(XV-X(J-1))/WJ
      D2=(X(J)-XV)/WJ
      D3=WJ*WJ/6.
      YV=D1*(Y(J)+D3*(D1*D1-1.)*S(J))
      YV=YV+D2*(Y(J-1)+D3*(D2*D2-1.)*S(J-1))
      RETURN
      ENTRY GRATA(XV,YV,N)
      DO 9 JJ=2,N
      J=JJ
      IF(XV.GT.X(J)) GOTO 9
      GOTO 10
    9 CONTINUE
   10 WJ=W(J)
      D1=((XV-X(J-1))/WJ)**2
      D2=((X(J)-XV)/WJ)**2
      D3=.0825*WJ*WJ
      YV=Z(J-1)+.5*WJ*(D1*(Y(J)+D3*(D1-2.)*S(J))+(1.-D2)*(Y(J-1)+D3*(D2
     *-1.)*S(J-1)))
      RETURN
      END
//GO.SYSIN DD *
EXAMPLE PROBLEM #9    P-68 AND P-107
ROCKY MOUNTAIN PETROLEUM CONSULTANTS          SALT LAKE CITY    UTAH
    2.4E+7    4964.98  4.6897  360.
  .209 .205 275.      3.0E-6
3699.1.538       6.8E-6   .3
  .200 .03
14
    0.3793.1.533..000835      900.                  .25.0190 900.
   91.3788.1.535..000840      900.        0.        .25.0180 900.
  182.3774.1.536..000840      900.        370.      .25.0180 900.
  273.3748.1.537..000846      900.        1030.     .25.0175 900.
  365.3709.1.538..000853      900.        1750.     .30.0173 900.
  456.3680.1.543..000861      900.        2834.     .35.0172 890.
  547.3643.1.548..000870      919.        4840.     .40.0171 887.
  638.3595.1.556..000882      914.        7749.     .45.0170 885.
  730.3547.1.563..000892      910.        13895.    .50.0169 875.
  821.3518.1.568..000899      911.        24808.    .55.0168 870.
  912.3485.1.573..000907      917.        37653.    .60.0167 865.
 1003.3437.1.578..000915      937.        58449.    .65.0166 855.
 1095.3416.1.594..000923      952.        111863.   .75.0165 840.
 1186.3379.1.591..000933      970.        163250.   .85.0164 820.
20
0.        0.
15.       9.949
30.       16.742
60.       28.691
75.       34.247
90.       39.626
105.      44.858
120.      49.968
135.      54.976
```

```
150.        59.895
165.        64.737
180.        69.512
195.        74.226
210.        78.886
225.        83.497
240.        88.062
255.        92.589
270.        97.081
285.        101.540
300.        105.968
15
.5000       .9999
.6340       .6700
.6680       .3270
.6940       .2400
.7180       .1830
.7240       .1400
.7680       .1020
.7980       .1020
.8270       .0510
.8590       .0390
.8930       .0200
.9200       .0165
.9600       .0043
.9940       .0000
.9990       .0000
/*
//
//LRPMGT12 JOB (U0290),PADGETT
/*JOBPARM F=BACK
// EXEC FORTGCLG
//FORT.SYSIN DD *
```

## OUTPUT LISTING

```
ESTIMATION OF FUTURE PERFORMANCE FOR WATER DRIVE
RESERVOIR USING UNSTEADY-STATE WATER INFLUX EQUATION

EXAMPLE PROBLEM #9    P-68 AND P-107
ROCKY MOUNTAIN PETROLEUM CONSULTANTS         SALT LAKE CITY    UTAH

ORIGINAL OIL IN PLACE       24000000. STB
```

| TIME | ESTIMATED RES.PRESS | CUM.OIL PROD | INSTI.GOR | CUM.WATER PRODUCED | CUM.WATER INFLUX |
|------|------|------|------|------|------|
| DAYS | PSIA | BBL | SCF/STD | BBL | BBL |
| 91. | 3788. | 34250. | 900. | 0. | 4577. |
| 182. | 3774. | 62974. | 900. | 370. | 25092. |
| 273. | 3748. | 111478. | 900. | 1030. | 76378. |
| 365. | 3709. | 190173. | 901. | 1750. | 174234. |
| 456. | 3680. | 356173. | 913. | 2834. | 312404. |
| 547. | 3643. | 537451. | 935. | 4840. | 485690. |
| 638. | 3595. | 773843. | 1005. | 7749. | 710491. |
| 730. | 3547. | 959775. | 1181. | 13895. | 990024. |
| 821. | 3518. | 1100165. | 1451. | 24808. | 1301545. |
| 912. | 3485. | 1181272. | 1709. | 37653. | 1635164. |
| 1003. | 3437. | 1267734. | 1919. | 58449. | 2012431. |
| 1095. | 3416. | 1430128. | 2225. | 111863. | 2416782. |
| 1186. | 3379. | 1430128. | 2519. | 163250. | 2840375. |

**BIBLIOGRAPHY**

Craft, B. C., and Hawkins, M. F.: *Applied Petroleum Reservoir Engineering,* Prentice-Hall, Englewood Cliffs, N.J. (1959).

Guerrero, E. T.: *Practical Reservoir Engineering,* The Petroleum Publishing Company, Tulsa (1968).

Hurst, W.: "Water Influx into a Reservoir and its Application to the Equation of Volumetric Balance," *Trans.,* AIME (1943).

Slider, H. C.: *Practical Petroleum Reservoir Engineering Methods,* The Petroleum Publishing Company, Tulsa (1976).

# 11    DISPERSED GAS INJECTION PERFORMANCE

**PURPOSE**

The program computes the dispersed gas injection performance of a solution gas drive reservoir as a function of reservoir pressure, with conformance taken into account. The computation is based on the Tracy form of the material balance equation.

The conformance factor $e$ is the fraction of the total net reservoir pore volume that is assumed to be contacted by the injection gas. The injection gas is assumed to be dispersed evenly throughout the conformable portion of the reservoir, and no frontal displacement occurs. Also, gas-oil miscibility is not considered. The production mechanism description can be divided into two regions—namely, conformable and nonconformable portions of the reservoir. The oil from the nonconformable portion of the reservoir is produced by only a depletion or a solution gas drive mechanism and flows into the conformable portion of the reservoir. In the conformable portion of the reservoir, a combination drive—that is, a solution gas drive and dispersed gas injection drive—is envisioned.

The volume of gas injected is provided by the user as a fraction or multiple of the produced gas volume. The series of pressure steps at which the performances are to be calculated as well as the respective fluid properties are provided by the user.

A major problem, however, for the user is to estimate the conformance factor and to determine if the oil from the nonconformable portion of the reservoir is flowing directly into the producing wells. Therefore, selection of these parameters will need considerable prudence and judgement.

The program provides a separate check by computing the solution gas drive, dispersed gas drive, and combination drive indexes. For the performance computation to be valid, the combination drive index must be close to unity.

113

## METHOD

$$\phi_o = \frac{B_o/B_g - R_s}{(B_o/B_g - R_s) - (B_{oi}/B_g - R_{si})}$$

$$\phi_g = \frac{1}{(B_o/B_g - R_s) - (B_{oi}/B_g - R_{si})}$$

$$\Delta N_P = \Delta N_{pe} + \Delta N_p^*(1 - e) = \frac{1 - N_{p((n-1)}\phi_o - G_{p(n-1)}\phi_g}{\phi_o + (1 - I)\left(\dfrac{R_{I(n-1)} + R_{Ie}}{2}\right)\phi_g}$$

$$SL_e = \frac{S_w + (1 - S_w)\{e - [N_p - N_p^*(1 - e)]\}B_o}{eB_{oi}}$$

$$R_{Ie} = R_s + (k_g/k_o)_e(\mu_o/\mu_g)(B_o/B_g)$$

$$N_p\phi_o + G_p\phi_g = 1$$

where

$\Delta N_p$ = Incremental oil recovery as a fraction of the original oil in place during the pressure decrement from $P_{n-1}$ to $P_n$, psi;

$\Delta N_{pe}$ = Incremental oil recovery as a fraction of the original oil in place from the conformable position of the reservoir during the pressure decrement from $P_{n-1}$ to $P_n$, psi;

$\Delta N_p^*$ = Incremental oil recovery from the nonconformable portion of the reservoir (only by depletion drive) as a fraction of the original oil in place during the pressure decrement from $P_{n-1}$ to $P_n$, psi;

$N_{P(n-1)}$ = Cumulative oil production as a fraction of the original oil-in-place to the end of pressure decrement to $P_{n-1}$, psi;

$G_{P(n-1)}$ = Cumulative gas production as a fraction of the original oil in place to the end of pressure decrement to $P_{n-1}$, psi;

$B_{oi}$ = Initial oil formation volume factor, bbl/STB;

$B_o$ = Oil formation volume factor, bbl/STB;

$B_g$ = Gas formation volume factor, bbl/SCF;

$R_s$ = Solution gas-oil ratio, SCF/STB;

$R_{si}$ = Initial solution gas-oil ratio, SCF/STB;

$RI_e$ = Instantaneous gas-oil ratio at $P_{n-1}$, SCF/STB;

$R_{I(n-1)}$ = Instantaneous gas-oil ratio at $P_{n-1}$, SCF/STB;

$e$ = Conformance factor;

$I$ = Gas injection volume, fraction of gas produced;

$N_p^*$ = Cumulative oil produced by depletion drive only, fraction of original oil in place;

$SL_e$ = Total liquid saturation in the reservoir, fraction;

$\mu_o$ = Oil viscosity, cp;
$\mu_g$ = Gas viscosity, cp;
$N_p$ = Cumulative oil produced, fraction of OOIP;
$G_p$ = Cumulative gas produced, fraction of OOIP;
$k_g/k_o$ = Gas-oil relative permeability ratio.

## PROGRAM DESCRIPTION
The program consists of a FORTRAN main program and a FORTRAN subprogram. Subprogram SPLINE is a curve-fitting program code utilized to generate continuous values of gas-oil relative permeability ratio values as a function of total liquid saturation from a set of user-supplied discrete values.

The main program uses an iterative procedure at each pressure step for solution convergence. The user must therefore provide a set of starting instantaneous gas-oil ratio values. These values are selected based on the user's judgment and experience.

### Input Format

| Card | Format | Content |
| --- | --- | --- |
| 1 | 20A4 | A(I) |
| 2 | 20A4 | B(I) |
| 3 | 3F10.0, F5.0, 2F5.3, F5.3 | OOIP, COP, CGP, RSI, BOI, SWI, PRINJ, CONF |
| 4 | 2I2 | IOPT, N |
| 5–N | 2F5.0, F5.3, F6.5, F4.2, F5.4, F5.0, F5.3 | C1(I), C3(I), C2(I), C4(I), C5(I), C37(I), C9(I), C38(I) |
| 6 (Needed only if IOPT = 1) | I2 | NN |
| 7–NN (Needed only if IOPT = 1) | 2F10.0 | SL(I), KGKO(I) |

### Description of Input Parameters

A(I) = Uses up to 80 alphanumeric characters for project identification;
B(I) = Uses up to 80 alphanumeric characters for project identification;
OOIP = Original oil in place, STB;
COP = Cumulative oil produced, fraction of OOIP;
CGP = Cumulative gas produced, fraction of OOIP;

RSI = Initial solution gas-oil ratio, SCF/STB;
BOI = Initial oil formation volume factor, bbl/STB;
SWI = Initial interstitial water saturation, fraction;
PRINJ = Pressure at which gas injection begins, psi;
CONF = Conformance factor, $0.0 \leq e \leq 1.0$;
IOPT = Relative permeability data-handling option:
IOPT = 1, gas-oil relative permeability data set is provided by the user; IOPT = 0, standard correlation for sandstone reservoir is used;
N = Number of pressure and fluid property data points provided;
C1(I) = Reservoir pressure, psi;
C2(I) = Oil formation volume factor, bbl/STB;
C3(I) = Solution gas-oil ratio, SCF/STB;
C4(I) = Gas formation volume factor, bbl/SCF;
C5(I) = Oil viscosity, cp;
C37(I) = Gas viscosity, cp;
C9(I) = Starting values of instantaneous gas-oil ratio, best guess values depending upon user judgment, SCF/STB;
C38(I) = Fraction of produced gas injected;
NN = Number of discrete liquid saturation–relative permeability data provided;
SL(I) = Total liquid saturation in the reservoir, fraction;
$K_G K_O$(I) = Corresponding relative permeability ratio.

**EXAMPLE PROBLEM**

Pressure and fluid property data for a depletion-type reservoir are given in table 11.1. The experimental values of total liquid saturation and the corresponding relative permeability ratios for the same reservoir are shown in table 11.2.

**Table 11.1 Pressure and Fluid Property Data for a Depletion-Type Reservoir**

| Pressure (psi) | $B_o$ (bbl/STB) | $R_s$ (SCF/STB) | $B_g$ (bbl/SCF) | $\mu_o$ (cp) | $\mu_g$ (cp) | $R_I$ (estimate) (SCF/STB) | Gas Injection (fraction of produced gas) |
|---|---|---|---|---|---|---|---|
| 1700 | 1.265 | 540 | 0.00132 | 1.19 | 0.02 | 540 | 0 |
| 1500 | 1.241 | 490 | 0.00150 | 1.22 | 0.02 | 710 | 0 |
| 1300 | 1.214 | 440 | 0.00175 | 1.25 | 0.02 | 1150 | 0 |
| 1100 | 1.191 | 387 | 0.00210 | 1.30 | 0.02 | 1470 | 0 |
| 900 | 1.161 | 334 | 0.00262 | 1.35 | 0.02 | 1900 | 0.6 |
| 700 | 1.147 | 278 | 0.00344 | 1.50 | 0.02 | 2000 | 0.6 |
| 500 | 1.117 | 220 | 0.00495 | 1.80 | 0.02 | 2500 | 0.6 |
| 300 | 1.093 | 160 | 0.00860 | 2.28 | 0.02 | 2600 | 0.6 |
| 100 | 1.058 | 84 | 0.02720 | 3.22 | 0.02 | 1850 | 0.6 |

**Table 11.2 Liquid
Saturation versus
Relative Permeability
Ratio**

| $S_L$ | $k_g/k_o$ |
|-------|-----------|
| 0.634 | 0.670 |
| 0.681 | 0.375 |
| 0.706 | 0.280 |
| 0.722 | 0.225 |
| 0.748 | 0.167 |
| 0.775 | 0.118 |
| 0.779 | 0.113 |
| 0.815 | 0.073 |
| 0.846 | 0.0495 |
| 0.883 | 0.0300 |
| 0.920 | 0.0165 |
| 0.960 | 0.0043 |

Find the performance of the reservoir, if dispersed gas injection is initiated when the reservoir pressure falls below 900 psia and when 60% of the produced gas is injected into the formation. Other available information is as follows:

Original oil in place = 500 MMSTB,
Initial formation volume factor = 1.265 bbl/STB,
Interstitial water saturation = 25%,
No water production,
Reservoir conformance factor = 50%.

# INPUT LISTING

| STATEMENT NUMBER | 1,2,3,4,5 | 6 | 7,8,9,10 | 11,12,13,14,15,16,17,18,19,20 | 21,22,23,24,25,26,27,28,29,30 | 31,32,33,34,35,36,37,38,39,40 | 41,42,43,44,45,46,47,48,49,50 | 51,52,53,54,55,56,57,58,59 |
|---|---|---|---|---|---|---|---|---|

Legend:
0 = ZERO   Ø = ALPHA O
1 = ONE    I = ALPHA I
2 = TWO    Z = ALPHA Z

```
PRØBLEM #110      ANU PØØL      SALTLAKE CITY      UTAH
RØTHUMØRE ESTATIG      P.817      P.1133
5.00.00.0      0.0
1.19
1.70.0.  5.4.0.  1.26.5.001.13.21.119..02   Ø.5.4.0..0.0
15.0.0.  4.9.0.  1.24.1.001.150.122..02      7.10..0.0
13.0.0.  4.4.0.  1.214.001.75.125..02        1.150.0.0.0
1.10.0.  3.8.7.  1.19.1.002.10.130..02       1.417.0.0.0
9.0.0.   3.34.   1.16.1.002.62.135..02       1.910.0.0.6
7.0.0.   27.8.   1.14.7.003.41.150..02       2.000.0.0.6
5.0.0.   2.20.   1.117.004.95.180..02        2.500.0.0.6
3.0.0.   16.0.   1.093.008.62.128..02        2.600.0.0.6
1.0.0.   8.4.    1.058.007.20.132..02        1.850.0.0.6
1.2
.6.34    .6.17.0
.6.81    .3.75
.7.06    .2.8.0
.7.22    .2.2.5
.7.48    .1.6.7
.7.75    .1.1.8
.7.79    .1.1.3
.8.15    .0.7.3
.8.46    .0.4.9.5
.8.83    .0.3.0.0
.9.2.0   .0.1.6.5
.9.6.0   .0.0.4.3
```

## SOURCE LISTING

```
C       THIS PROGRAM IS DESIGNED TO CALCULATE DISPARSED INJECTION
C       PERFORMANCE FOR A SOLUTION GAS DRIVE(UNDER SATURATED) RESERVOIR
C       USING TRACY FORM OF MATERIAL BALANCE EQUATIONS TAKING CONFORMANCE
C       INTO ACCOUNT.   INJECTED GAS VOLUMES ARE PROVIDED AS A FRACTION OF
C       GAS PRODUCED.  NO MISCIBILITY IS ASSUMED.  RESERVOIR FLUID AND
C       ROCK PROPERTY DATA AS WELL AS INITIAL OIL IN PLACE VALUES ARE TO
C       BE PROVIDED OR OPTIONALLY STANDARD CORRELATTION FOR SANDSTONE
C       RESERVOIR CAN BE USED.
C
        REAL KGKO
        DIMENSION C1(20),C2(20),C3(20),C4(20),C5(20),C6(20),C7(20),C8(20)
       *,C9(20),C10(20),C11(20),C12(20),C13(20),C14(20),C15(20),C16(20),
       *C17(20),C18(20),C19(20),C20(20),C21(20),C22(20),C23(20),C24(20),
       *C25(20),C26(20),C27(20),C28(20),C29(20),C30(20),C31(20),C32(20)
       *,C33(20),C34(20),C35(20),C36(20),C37(20),C38(20),C39(20),C40(20)
       *,A(20),B(20),SL(20),KGKO(20)
       *,C6D(20),C7D(20),C8D(20),C9D(20),C10D(20),C11D(20),
       *C12D(20),C13D(20),C14D(20),C15D(20),C16D(20),C17D(20),
       *C18D(20),C19D(20),C21D(20),C22D(20),C23D(20),C24D(20),
       *C25D(20),C26D(20),C27D(20),C28D(20),C29D(20),C30D(20),
       *C31D(20),C32D(20),C33D(20),C34D(20),C35D(20),C36D(20),
       *C37D(20),C20D(20)
C
C       C38=FRACTION OF PRODUCED GAS INJECTED
C       OOIP=ORIGINAL OIL IN PLACE, STB
C       SWI=WATER SATURATION, FRACTION
C       BOI=INITIAL FORMATION VOLUME FACTOR, RES.BBL/STB
C
C
C       RSI=INITIAL SOLUTION GAS-SATURATION, FRACTION
C       COP=CUM OIL.PRODUCTION AT INITIAL COMPUTATIONAL STEP
C       CGP=CUM.GAS.PRODUCTION AT STARTING POINT OF CALCULATION
C       PRINJ=PRESSURE AT WHICH GAS INJECTION IS STARTED, PSI
C       CONF=CONFORMANCE FACTOR, FRACTIONAL
C       IOPT=RELATIVE PERMEABILITY DATA HANDLING OPTION
C            1, SATURATICN-RELATIVE PERMEABILITY INFORMATION PROVIDED
C            0, STANDARD RELATIVE PERMEABILITY CORRELATION ARE USED
C
C       C37=GAS VISCOSITIES
C       C9=ASSUMED/STARTING VALUES OF INST.GOR AS FIRST GUESS, SCF/STB
C       SL=TOTAL LIQUID SATURATION, FRACTION
C       KGKO=GAS-OIL RELATIVE PERMEABILITY
C       NUMBER OF PRESSURE POINTS AT WHICH CALCULATIONS ARE TO BE PER-
C       FORMED.  INCLUDING THE STARTING OR INITIAL CONDITION.
C       NN=NUMBER OF RELATIVE PERMEABILITY-SATURATION POINT DATA PROVIDED
C       N=NUMBER OF PRESSURE POINTS FOR CALCULATIONS
C
        READ(5,1)(A(I),I=1,20)
        READ(5,1)(B(I),I=1,20)
      1 FORMAT(20A4)
        READ(5,2)OOIP,COP,CGP,RSI,BOI,SWI,PRINJ,CONF
      2 FORMAT(3F10.0,F5.0,2F5.3,F5.0,F5.3)
        READ(5,3)IOPT,N
      3 FORMAT(2I2)
        DO 100 I=1,N
        READ(5,4)C1(I),C3(I),C2(I),C4(I),C5(I),C37(I),C9(I),C38(I)
      4 FORMAT(2F5.0,F5.3,F6.5,F4.2,F5.4,F5.0,F5.3)
        X=(((C2(I)/C4(I))-C3(I))-((BOI/C4(I))-RSI))
        IF(I.GT.1)C6(I)=((C2(I)/C4(I))-C3(I))/X
        IF(I.GT.1)C7(I)=1./X
    100 CONTINUE
        IF(IOPT.EQ.0)GO TO 150
        READ(5,5)NN
      5 FORMAT(I2)
        DO 105 I=1,NN
        READ(5,6)SL(I),KGKO(I)
      6 FORMAT(2F10.9)
    105 CONTINUE
    150 CONTINUE
        C33(1)=CGP
        C18(1)=COP
        C22D(1)=C18(1)
        C13D(1)=C9(1)
        C30D(1)=C3(1)
        DO 2080 I=2,N
        IKNT=0
        C6D(I)=C2(I)/C4(I)
        C8(I)=C6D(I)
        C7D(I)=C6D(I)-C3(I)
        C8D(I)=BOI/C4(I)
```

```
      C9D(I)=C8D(I)-RSI
      C10D(I)=C7D(I)-C9D(I)
      C11D(I)=C7D(I)/C10D(I)
      C12D(I)=1./C10D(I)
      C13D(I)=C9(I)
 2078 CONTINUE
      C14D(I)=0.5*(C13D(I-1)+C13D(I))
      C15D(I)=C22D(I-1)*C11D(I)
      C16D(I)=C33D(I-1)*C12D(I)
      C17D(I)=C15D(I)+C16D(I)
      C18D(I)=1.-C17D(I)
      C19D(I)=C14D(I)*C12D(I)
      C20D(I)=C11D(I)+C19D(I)
      C21D(I)=C18D(I)/C20D(I)
      C22D(I)=C22D(I-1)+C21D(I)
      C19(I)=C22D(I)
      C23D(I)=1.-C22D(I)
      C24D(I)=C2(I)*C23D(I)
      C25D(I)=(1.-SWI)*C24D(I)/BOI
      C26D(I)=SWI+C25D(I)
      IF(IOPT.EQ.0)GO TO 2076
      CALL SPLINE(SL,KGKO,NN)
      CALL GRATA(C26D(I),AA,NN)
      CALL TERPA(C26D(I),TX,NN)

      C27D(I)=TX
      GO TO 2077
 2076 CONTINUE
      SSTAR=C25D(I)/(1.-SWI)
      C27D(I)=(((1.-SSTAR)**2)*(1.-SSTAR**2))/(SSTAR**4)
 2077 CONTINUE
      C28D(I)=(C5(I)/C37(I))*C6D(I)
      C29D(I)=C27D(I)*C28D(I)
      C30D(I)=C3(I)+C29D(I)
      C31D(I)=(C30D(I-1)+C30D(I))*0.5
      XTEST=ABS((C30D(I)-C13D(I))/C13D(I))
      IF(XTEST.LE.0.01)GO TO 2075
      C13D(I)=C30D(I)
      IKNT=IKNT+1
      IF(IKNT.GT.100)GO TO 2075
      GO TO 2078
 2075 CONTINUE
      C32D(I)=C21D(I)*C31D(I)
      C33D(I)=C33D(I-1)+C32D(I)
      C34D(I)=C33D(I)*C12D(I)

      C35D(I)=C22D(I)*C11D(I)
      C36D(I)=C34D(I)+C35D(I)
      C37D(I)=OOIP*C22D(I)
 2080 CONTINUE
      DO 2090 I=1,N
      II=I
      IF(C1(I).LT.PRINJ)GO TO 2091
 2090 CONTINUE
 2091 CONTINUE
      C18(II-1)=C19(II-1)
      C33(II-1)=C33D(II-1)
      C9(II-1)=C30D(II-1)
      C29(II-1)=C30D(II-1)
C     CALCULATIONS WITH GAS INJECTION
      DO 180 I=II,N
      IKNT=0
  178 CONTINUE
      C10(I)=0.5*(C9(I-1)+C9(I))
      C11(I)=C10(I)*C7(I)
      C12(I)=(1.-C38(I))*C11(I)
      C13(I)=C6(I)+C12(I)
      C14(I)=C18(I-1)*C6(I)
      C15(I)=C33(I-1)*C7(I)
      C16(I)=1.-C14(I)-C15(I)
      C17(I)=C16(I)/C13(I)
      C18(I)=C18(I-1)+C17(I)
      IF(C1(I).LT.PRINJ)C20(I)=C19(I)*(1.-CONF)
      C21(I)=C18(I)-C20(I)
      IF(C1(I).LT.PRINJ)C22(I)=C2(I)/(CONF*BOI)
      IF(C1(I).LT.PRINJ)C23(I)=CONF-C21(I)
      C24(I)=(1.-SWI)*C22(I)*C23(I)
      C25(I)=SWI+C24(I)
      IF(IOPT.EQ.0)GO TO 176
      CALL SPLINE(SL,KGKO,NN)
      CALL GRATA(C25(I),AA,NN)
      CALL TERPA(C25(I),TX,NN)
      C26(I)=TX
      GO TO 177

  176 CONTINUE
      SSTAR=C24(I)/(1.-SWI)
      C26(I)=(((1.-SSTAR)**2)*(1.-SSTAR**2))/(SSTAR**4)
```

```
177 CONTINUE
    C27(I)=C5(I)/C37(I)
    C28(I)=C26(I)*C27(I)*C8(I)
    C29(I)=C3(I)+C28(I)
    C30(I)=0.5*(C29(I-1)+C29(I))
    XTEST=ABS((C9(I)-C29(I))/C9(I))
    IF(XTEST.LE.0.01)GO TO 175
    C9(I)=C29(I)
    IKNT=IKNT+1
    IF(IKNT.GT.100)GO TO 180
    GO TO 178
175 CONTINUE
    C31(I)=C17(I)*C30(I)


    C32(I)=C31(I)*(1.-C38(I))
    C33(I)=C33(I-1)+C32(I)
    C34(I)=C33(I)*C7(I)
    C35(I)=C18(I)*C6(I)
    C36(I)=C34(I)+C35(I)
180 CONTINUE
    WRITE(6,7)
  7 FORMAT(//,3X,'RESERVOIR PERFORMANCE CALCULATIONS USING TRACY',
   *' FORM OF MBE')
    IF(PRINJ.GT.C1(N))WRITE(6,8)
  8 FORMAT(/,3X,'DISPERSED GAS INJECTION PERFORMANCE IS ',
   *'CONSIDERED')
    WRITE(6,9)(A(I),I=1,20)
    WRITE(6,9)(B(I),I=1,20)
  9 FORMAT(//,3X,20A4)
    WRITE(6,10)OOIP
 10 FORMAT(/,3X,'ORIGINAL OIL IN PLACE',3X,F15.0,1X,'STB')
    WRITE(6,11)COP
 11 FORMAT(/,3X,'CUM.OIL PRODUCED',3X,F19.0,1X,'STB')
    IF(PRINJ.GT.C1(N))WRITE(6,12)CONF
 12 FORMAT(/,3X,'ASSUMED CONFORMANCE FACTOR',3X,F5.3,1X,'RATIO')
    WRITE(6,13)
 13 FORMAT(//,3X,'PRESS.',3X,'FRAC.PROD',3X,'CUM.OIL PROD',
   *3X,'CUM.GAS PROD.',3X,'GDI',5X,'SDI',3X,'SDI+',3X,'INST.GOR',3X,
   *'CUM.GAS.INJ ED',3X,'PCT.REC')
    WRITE(6,14)
 14 FORMAT(4X,'PSIA',4X,'GAS INJECT',6X,'MMBBL',11X,'MMCF',21X,'GDI')
    DO 190 I=2,N
    IF(C1(I).LT.PRINJ)C40(I)=100*C18(I)
    IF(C1(I).GE.PRINJ)C40(I)=100*C22D(I)
    IF(C1(I).GE.PRINJ)C18(I)=C22D(I)*OOIP*1.E-6
    IF(C1(I).LT.PRINJ)C18(I)=C18(I)*OOIP*1.E-6
    IF(C1(I).GE.PRINJ)C33(I)=C33D(I)*OOIP*1.E-6
    IF(C1(I).LT.PRINJ)C33(I)=C33(I)*OOIP*1.E-6
    IF(C1(I).GE.PRINJ)C34(I)=C34D(I)
    IF(C1(I).GE.PRINJ)C35(I)=C35D(I)
    IF(C1(I).GE.PRINJ)C36(I)=C36D(I)
    IF(C1(I).GE.PRINJ)C30(I)=C31D(I)
    IF(C1(I).GE.PRINJ)C39(I)=0.
    IF(C1(I).LT.PRINJ)C39(I)=C39(I-1)+(C33(I)-C33(I-1))*C38(I)
    WRITE(6,15)C1(I),C38(I),C18(I),C33(I),C35(I),C34(I),C36(I),
   *C30(I),C39(I),C40(I)
 15 FORMAT(1X,F8.0,F7.1,F16.0,F16.0,F10.2,F8.2,F7.2,
   *F11.0,F17.3,F10.3)
190 CONTINUE
    END
    SUBROUTINE SPLINE(XI,YI,N)
    DIMENSION W(99),Q(99),B(99),A(99),C(99),S(99),Z(99)
    DIMENSION XI(50),YI(50),X(99),Y(99)
    DATA A(1),C(1),Z(1)/-1.0,0.,0./
    DO 11 I=1,N
    X(I)=XI(I)
    Y(I)=YI(I)
 11 CONTINUE
    DO 1 J=2,N
    W(J)=X(J)-X(J-1)
  1 CONTINUE
    NM=N-1
    DO 2 J=2,NM
    WJ=W(J)
160 WP=W(J+1)
    WS=WJ+WP
    QJ=WJ/WS
    Q(J)=QJ
    QA=1.-.5*QJ*A(J-1)
    A(J)=.5*(1.-QJ)/QA
    B(J)=3.*(WP*Y(J-1)-WS*Y(J)+WJ*Y(J+1))/WP/WJ/WS
    C(J)=(B(J)-.5*QJ*C(J-1))/QA
  2 CONTINUE
    S(N)=C(NM)/(1.+A(NM))
    S(NM)=S(N)
    NMM=N-2
```

```
          DO 3 JJ=1,NMM
          J=NMM-JJ+1
          S(J)=C(J)-A(J)*S(J+1)




     3 CONTINUE
          DO 4 J=2,N
          Z(J)=Z(J-1)+.5*W(J)*(Y(J)+Y(J-1)-.0825*W(J)**2*(S(J)+S(J-1)))
     4 CONTINUE
          RETURN
C
          ENTRY DERIVA(XV,YV,N)
          DO 5 JJ=2,N
          J=JJ
          IF(XV.GT.X(J))GOTO 5
          GOTO 6
     5 CONTINUE
     6 WJ=W(J)
          D1=((XV-X(J-1))/WJ
          D2=(X(J)-XV)/WJ
          YV=(Y(J)-Y(J-1))/WJ+.5*WJ*((D1*D1-.333333)*S(J)
     *-(D2*D2-.333333)*S(J-1))
          RETURN
          ENTRY TERPA(XV,YV,N)
          DO 7 JJ=2,N
          J=JJ
          IF(XV.GT.X(J))  GOTO 7
          GOTO 8
     7 CONTINUE
     8 WJ=W(J)
          D1=((XV-X(J-1))/WJ
          D2=(X(J)-XV)/WJ
          D3=WJ*WJ/6.
          YV=D1*(Y(J)+D3*(D1*D1-1.)*S(J))
          YV=YV+D2*(Y(J-1)+D3*(D2*D2-1.)*S(J-1))
          RETURN
          ENTRY GRATA(XV,YV,N)
          DO 9 JJ=2,N
          J=JJ
          IF(XV.GT.X(J))  GOTO 9
          GOTO 10
     9 CONTINUE
    10 WJ=W(J)
          D1=((XV-X(J-1))/WJ)**2
          D2=((X(J)-XV)/WJ)**2
          D3=.0825*WJ*WJ
          YV=Z(J-1)+.5*WJ*(D1*(Y(J)+D3*(D1-2.)*S(J))+(1.-D2)*(Y(J-1)+D3*(D2
     *-1.)*S(J-1)))
          RETURN
          END
```

```
PROBLEM # 10               ANU POOL      SALT LAKE CITY   UTAH
ROTHMORE ESTATE     P 87   P 133
      5.0E8       0.000       0.000540. 1.256.250 900. .500
  1 9
1700.  540.1.265.001321.19.02      540.0.0
1500.  490.1.241.001501.22.02      710.0.0
1300.  440.1.214.001751.25.02     1150.0.0
1100.  387.1.191.002101.30.02     1470.0.0
 900.  334.1.161.002621.35.02     1900.0.0
 700.  278.1.147.003441.50.02     2000.0.6
 500.  220.1.117.004951.80.02     2500.0.6
 300.  160.1.093.008602.28.02     2600.0.6
 100.   84.1.058.027203.22.02     1850.0.6
12
.634          .670
.681          .375
.706          .28
.722          .225
.748          .1670
.775          .118
.779          .1130
.815          .0730
.846          .0495
.883          .03
.920          .0165
.960          .0043
/*
//
```

# OUTPUT LISTING

RESERVOIR PERFORMANCE CALCULATIONS USING TRACY FORM OF MBE

DISPERSED GAS INJECTION PERFORMANCE IS CONSIDERED

PROBLEM # 10        ANU POOL        SALT LAKE CITY    UTAH

ROTHMORE ESTATE    P 87   P 133

ORIGINAL OIL IN PLACE        500000000. STB

CUM. OIL PRODUCED            0. STB

ASSUMED CONFORMANCE FACTOR   0.500 RATIO

| PRESS. PSIA | FRAC. PROD GAS INJECT | CUM. OIL PROD MMBBL | CUM. GAS PROD. MMCF | GDI | SDI | SDI+ GDI | INST. GOR | CUM. GAS. INJ ED | PCT. REC |
|---|---|---|---|---|---|---|---|---|---|
| 1500. | 0.0 | 21. | 12922. | 0.35 | 0.65 | 1.00 | 622. | 0.0 | 4.157 |
| 1300. | 0.0 | 38. | 24411. | 0.25 | 0.75 | 1.00 | 926. | 0.0 | 7.503 |
| 1100. | 0.0 | 55. | 51121. | 0.16 | 0.84 | 1.00 | 1318. | 0.0 | 10.949 |
| 900. | 0.0 | 71. | 72231. | 0.09 | 0.91 | 1.00 | 1656. | 0.0 | 14.103 |
| 700. | 0.6 | 102. | 109651. | 0.05 | 0.95 | 1.00 | 2542. | 19452.187 | 20.479 |
| 500. | 0.6 | 125. | 145522. | 0.00 | 1.00 | 1.00 | 4942. | 40974.824 | 24.916 |
| 300. | 0.6 | 144. | 185587. | -0.03 | 1.03 | 1.00 | 5199. | 65013.484 | 28.769 |
| 100. | 0.6 | 167. | 232094. | -0.03 | 1.03 | 1.00 | 4920. | 92918.125 | 33.496 |

**BIBLIOGRAPHY**
Guerrero, E. T.: *Practical Reservoir Engineering*, The Petroleum Publishing Company, Tulsa (1968).

# III ENHANCED OIL RECOVERY AND PERFORMANCE BY EMPIRICAL METHODS

# 12    IN SITU COMBUSTION PERFORMANCE USING THE OIL-DISPLACED/VOLUME-BURNED METHOD

## PURPOSE

The program essentially performs a set of computations using the Fassihi, Gobran, and Ramey algorithm for the oil-displaced/volume-burned method for in situ performance predictions. These authors claim that the method can be used to provide accurate engineering and economic evaluation for designing and monitoring in situ combustion projects.

The program provides estimates of oil recovery, oil rates, and air-fuel ratios. The user has the option to provide either fuel concentration values or to use values computed from data obtained from combustion tube runs and combustion chemistry. The combustion tube run data are also needed for completing the performance calculations.

## METHOD

The amount of initial oil and gas saturations is determined by using conventional well-logging, coring, and material balance or even tracer techniques. Displaced oil is equal to the initial oil minus the final oil minus the burned oil. Fuel concentration is a very important parameter and is influenced by fluid properties, formation lithology, and operating conditions. The volume of air required to burn a unit weight of fuel is determined utilizing flue gas composition data available from the laboratory combustion tube runs.

The computed values of air requirement, fuel concentration, initial oil and gas saturation, and required oxygen are used in conjunction with the oil-displaced/volume-burned correlations. The oil-displaced/volume-burned correlations (empirical) are:

$$X = \frac{V_B - V_{B(O)}}{100 - V_{B(O)}}$$

$$V_{B(O)} = 0.14714 S_g + 0.01071 S_g^2$$

$$MD = \text{Maximum deviation} = 26.8229 - 0.4678 S_g$$

$$Y = \frac{\text{Deviation}}{\text{Maximum deviation}} = 6.7752X - 15.9478X^2 + 16.1872X^3 - 7.0146X^4$$

$$R = \text{Ultimate recovery} = S_{oi} - B$$

$$\text{Slope} = \frac{100}{100 - V_{B(O)}} + \frac{MD}{100 - V_{B(O)}} \frac{dY}{dX}$$

$$\text{Oil recovery (percent)} = 100X + Y \times MD$$

Oil recovery (STB) = {Oil recovery (percent) × ultimate recovery,
R(STB/acre-ft) × pattern area (acre)
× reservoir thickness}/100

where

$V_B$ = Volume burned, percent of bulk volume;
$V_{B(O)}$ = Volume burned at oil breakthrough, percent;
$S_g$ = Reservoir gas saturation;
$B$ = Fuel consumed, bbl/acre-ft;
$S_{oi}$ = Initial oil saturation, bbl/acre-ft;
$dY/dX$ = Slope of the oil-displaced–volume-burned plot.

## PROGRAM DESCRIPTION

The program consists of a simple FORTRAN main program that computes the entire volume burned (100%) in 10 steps. This assumes that at each step 10% of the bulk volume will be burned. The user may provide a value for fuel concentration. Otherwise, the fuel concentration will be computed from laboratory combustion tube runs utilizing combustion chemistry. Under the first option (providing fuel concentration value), the user must also provide the air-fuel data.

### Input Format

| Card | Format | Content |
| --- | --- | --- |
| 1 | 20A4 | A(J) |
| 2 | 20A4 | B(I) |
| 3 | 4F6.2, F11.2, F5.2, F10.2 | XN2, CO2, CO, O2, VF, RT, QG |
| 4 | 5F10.2 | AREA, H, SOI, SG, Q |
| 5 | 2F10.2 | CF, AFR |

**Description of Input Parameters**

$A(J)$ = Uses up to 80 alphanumeric characters for project identification;
$B(I)$ = Uses up to 80 alphanumeric characters for project identification;
XN2 = Mol percentage of nitrogen in the flue gas (combustion tube run);
CO2 = Mol percentage of carbon dioxide in the flue gas (combustion tube run);
CO = Mol percentage of carbon monoxide in the flue gas (combustion tube run);
O2 = Mol percentage of oxygen in the flue gas (combustion tube run);
VF = Combustion tube front velocity, ft/hr;
RT = Combustion tube radius, ft;
QG = Combustion tube gas flow rate, SCF/hr;
AREA = Pattern area, acre;
H = Average reservoir thickness, ft;
SOI = Initial oil saturation, bbl/acre-ft;
SG = Gas saturation, percent;
Q = Field injection rate, MCF/day;
CF = Fuel concentration, lb/ft$^3$;
AFR = Air-to-fuel ratio, SCF/lb.

**EXAMPLE PROBLEM**

Determine in situ combustion for a reservoir with the following fluid and reservoir characteristics. The combustion tube experiment data are not available.

Oil saturation = 1540 bbl/acre-ft,
Reservoir gas saturation = 4%,
Pattern area = 1 acre,
Average reservoir thickness = 1 ft,
Air injection rate = 1 MCFD,
Fuel concentration = 2.1 lb/ft$^3$,
Air-fuel ratio = 184 SCF/lb.

**INPUT LISTING**

FORTRAN CODING FORM

UU/CC   NAME
        DATE

PROGRAM
ROUTINE

0 = ZERO        1 = ONE         2 = TWO
Ø = ALPHA O     I = ALPHA 1     Z = ALPHA Z

| STATEMENT NUMBER 1,2,3,4,5 | 6 | 7,8,9,10,11,12,13,14,15,16,17,18,19,20,21,22,23,24,25,26,27,28,29,30,31,32,33,34,35,36,37,38,39,40,41,42,43,44,45,46,47,48,49,50,51,52,53,54,55,56,57,58,59,60 |
|---|---|---|
| CARD | | |
| 1. | INSITU PRØJECT | DIWEEP GHØSH |
| 2. | SANGAS | CALIEFØRNIA |
| 3. | 01.00 | 01.000 | 0.000 | 01.010 01.000 | 0.00 |
| 4. | 1.0 | 11.0 | 15.40. | 41.0 | 1.0 |
| 5. | 2.1 | 184.0 | | |

## SOURCE LISTING

```
C       THIS PROGRAM CALCULATES IN SITU COMBUSTION PERFORMANCE USING
C       FASSIHI,GOBRAN,RAMEY, ALGORITHM OIL DISPLACED/VOLUME BURNED
        DIMENSION A(20),B(20)
        READ(5,3)(A(I),I=1,20)
        READ(5,3)(B(I),I=1,20)
      3 FORMAT(20A4)

        READ(5,5)XN2,CO2,CO,O2,VF,RT,QG
        READ(5,6)AREA,H,SOI,SG,Q
      5 FORMAT(4F6.2,F11.2,F5.2,F10.2)
      6 FORMAT(5F10.2)
        READ(5,9)CF,AFR
        WRITE(6,1)
      1 FORMAT(//,3X,' IN SITU COMBUSTION PERFORMANCE BY VOLUME BURNED METH
       *OD',/,3X,'FASSIHI,GOBRAN,RANEY ALGORITHM')
        WRITE(6,2)(A(I),I=1,20)
        WRITE(6,2)(B(I),I=1,20)
      2 FORMAT(/,20X,20A4)
        WRITE(6,7)
        WRITE(6,8)
        IF(CF.GT.0.0) GO TO 11
C
C       XN2=MOL % NITROGEN IN THE FLUE GAS FROM COMBUSTION TUBE RUN
C       CO2=MOL % CARBON DIOXIDE IN THE COMBUSTION TUBE RUN FLUE GAS
C       CO=MOL % CARBON MONOXIDE IN THE COMBUSTION TUBE RUN FLUE GAS
C       O2=MOL % OXYGEN IN THE COMBUSTION TUBE RUN GAS
C       VF=COMBUSTION TUBE FRONT VELOCITY, FT/HR
C       RT=COMBUSTION TUBE RADIUS, FT
C       QG=COMBUSTION TUBE GAS FLOW RATE, SCF/HR
C       AREA=PATTERN AREA, AC
C       SOI=FIELD OIL SATURATION, BBL/AC-FT
C       H=NET PAY, FT
C       SG=GAS SATURATION IN THE RESERVOIR, %
C       Q=INJECTION RATE, MCF/DAY
C       CF=FUEL CONCENTRATION, LB/CF OF ROC
C       AFR=AIR FUEL RATIO MCF/LB
        HC=4.*(0.2658*XN2-CO2-O2-0.5*CO)/(CO2+CO)
        CF=(0.00129*QG*(CO2+CO)*(12.+HC))/(VF*RT**2)
        AFR=479.7*XN2/((CO2+CO)*(12.+HC))
     11 CONTINUE
        ASR=AFR*CF*43.56
        BB=CF*124.4
        R=SOI-BB
        VB=0.
        VBO=0.147143*SG+.010714*SG**2
        DO 100 I=1,10
        VB=VB+10.
        X=(VB-VBO)/(100.-VBO)
        XMD=26.82295-0.46787*SG
        Y=6.775267*X-15.947794*X**2+16.187*X**3-7.014659*X**4
        DYDX=6.775267-31.895588*X+48.561561*X**2-28.058636*X**3
        IF(Y.LT.0.0)GO TO 105
        IF(DYDX.LT.0.0)GO TO 105
        SLOPE=100./(100.-VBO)+(XMD/(100.-VBO))*DYDX
        CURAOR=ASR/(SLOPE*R)
        XNP=100.*X*Y*XMD
        XN=XNP*R*AREA*H/100.
        AIRREQ=ASR*AREA*H*VB*.01
        CUMAOR=AIRREQ/XN
        TIME=AIRREQ/Q
        EXAIR=(0.9*XNP-15.85)*0.01
        TOTAOR=CURAOR*(1.+EXAIR)
        WRITE(6,101)VB,TIME,XNP,XN,CURAOR,AIRREQ,CUMAOR,TOTAOR
    101 FORMAT(F15.0,F10.3,F10.3,F13.0,F9.3,F10.0,3X,F7.2,1X,F9.3)
    100 CONTINUE
    105 CONTINUE
      7 FORMAT(//,5X,'VOL BURNED',5X,'TIME ',3X,'OIL REC',5X,'OIL REC.',2X
       *,'CUR AOR',3X,'AIR REQ',3X,'CUM AOR',1X,'TOTAL AOR')
      9 FORMAT(2F10.2)
      8 FORMAT(10X,'%',9X,'DAYS',7X,'%',10X,'BBL',5X,'MCF/BBL',5X,'MCF',5X
       *,'MCF/BBL',3X,'MCF/BBL')
        STOP
        END
//GO.SYSIN DD *
        IN SITU PROJECT   DWEEP GHOSH
        SAUGAS    CALIFORNIA
0.00      0.00    0.00    0.00       0.00 0.00       0.00
          1.00    1.00    1540.0     4.00 1.0
          2.1     184.0
```

# OUTPUT LISTING

IN SITU COMBUSTION PERFORMANCE BY VOLUME BURNED METHOD
FASSIHI,GOBRAN,RANEY ALGORITHM

IN SITU PROJECT    DWEEP GHOSH

SAUGAS    CALIFORNIA

| VOL BURNED % | TIME DAYS | OIL REC % | OIL REC. BBL | CUR AOR MCF/BBL | AIR REQ MCF | CUM AOR MCF/BBL | TOTAL AOR MCF/BBL |
|---|---|---|---|---|---|---|---|
| 10. | 1683.157 | 21.914 | 280. | 6.375 | 1683. | 6.01 | 6.622 |
| 20. | 3366.315 | 39.901 | 510. | 8.416 | 3366. | 6.60 | 10.105 |
| 30. | 5049.469 | 53.741 | 687. | 10.720 | 5049. | 7.35 | 14.206 |
| 40. | 6732.629 | 64.862 | 829. | 12.993 | 6733. | 8.12 | 18.518 |

**BIBLIOGRAPHY**
Fassihi, M. R., Gobran, B. D., and Ramey, H. J. Jr.: "Algorithm Calculated Performance of In situ Combustion," *Oil and Gas J.* (November 16, 1981) 90–97.

# 13　IN SITU COMBUSTION PERFORMANCE USING EMPIRICAL CORRELATIONS

**PURPOSE**

The program performs a preliminary estimate of oil recovery and performance under an in situ combustion process based on empirical correlations published in U.S. DOE report DOE/ET/12072-2. This is not a rigorous analysis by any means and is not intended to be. The results obtained are very preliminary in nature. Hence, the user must exercise due caution and diligence when using this program.

The model is based on correlations derived from the results of 14 experimental (field) projects. The oil recovery is assumed to be a function of oil saturation, thickness of the reservoir, oil viscosity, and air injection.

**METHOD**

$$F = -0.12 + 0.00262h + 0.000114k + 2.23S_o + 0.000242kh/\mu_o - 0.000189D - 0.0000652\mu_o$$

$$N_b = 132FhA$$

$$N = 7758AhS_o/B_o$$

$$X = A_iEO_2S_{orw}\phi/[N(1 - \phi)]$$

$$C = [0.427S_{orw} - 0.00135h + 2.196(1/\mu_o)^{0.25}]X$$

$$\text{Incremental recovery} = 47(1 - e^{-1.2C}) = \frac{N_p + N_b}{N} \times 100$$

where

$F$ = Fuel content, lb/ft burned volume;
$h$ = Net pay, ft;
$k$ = Permeability, md;
$S_o$ = Original oil saturation, fraction;
$\mu_o$ = Oil viscosity, cp;
$D$ = Depth, ft;
$A$ = Area, acre;
$B_o$ = Oil formation volume factor;
$A_i$ = Cumulative air injected, MMCF;
$E_{o_2}$ = Oxygen utilization, fraction;
$S_{orw}$ = Oil saturation at the start of combustion, fraction;
$\phi$ = Porosity, fraction;
$N$ = Original oil in place, STB;
$N_p$ = Oil produced, bbl;
$N_b$ = Oil burned, bbl.

This relationship is assumed to be valid until the fire front arrives at the production well or severe air breakthrough occurs. Recovery increases as the air injection is continued, but the recovery efficiency (that is, bbl/MCF air injected) decreases exponentially. Maximum recovery is limited to 47% of oil in place.

## PROGRAM DESCRIPTION
The program consists of only a FORTRAN main program and can perform computations for a maximum of 20 years. However, the program is designed to terminate automatically whenever the air-oil ratio becomes greater than 30 MMCF/bbl.

## Input Format

| Card | Format | Content |
|---|---|---|
| 1 | 20A4 | A(I) |
| 2 | 20A4 | B(I) |
| 3 | 6F12.3 | D, H, PERM, POR, SO, BOI |
| 4 | 6F12.3 | SORW, SPACE, AIR, EO2, VIS |

## Description of Input Parameters

A(I) = Uses up to 80 alphanumeric characters for project identification;
B(I) = Uses up to 80 alphanumeric characters for project identification;

$D$ = Reservoir depth, ft;
$H$ = Reservoir net pay, ft;
PERM = Average permeability, md;
POR = Average porosity, fraction;
SO = Initial oil saturation, fraction;
BOI = Initial oil formation volume factor at the start of the project, bbl/STB;
SORW = Reservoir oil saturation at the start of the combustion, fraction;
SPACE = Reservoir area extent or pattern areal extent, acre;
AIR = Annual air injection rate, MMCF/year;
EO = Oxygen utilization;
VIS = Oil viscosity, cp.

**EXAMPLE PROBLEM**

Determine in situ combustion performance and recoveries for a reservoir with the following properties:

Depth = 1035 ft,
Net pay = 19 ft,
Spacing = 10 acre,
Initial oil saturation = 71.8%,
Oil saturation before project startup = 69.3%,
Oil viscosity = 110 cp,
Average reservoir porosity = 34%,
Average reservoir permeability = 500 md,
Annual air injection rate = 190 MMCF/yr.

INPUT LISTING

```
                                          0 = ZERO          1 = ONE           2 = TWO
                                          Ø = ALPHA O       1 = ALPHA 1       Z = ALPHA Z

STATEMENT
NUMBER
   1,2,3,4,5 6 7,8,9,10 11,12,13,14,15,16,17,18,19,20 21,...,30 31,...,40 41,...,50 51,...,60 61,...,70

1. SUBMIT ALFIELD          MADHOMILITIAL RESERVOIR   UNIALS AND
2. HAMILTON COUNTY   PLAINSBORO       P.3.5.   PROJG9.   PRØG19.
3.    0.35.        19.01      5.0.01.01   0..3.4.01   0..17.18   1..1.10
4.    01.69.3.     1.0..01    1.900.0.01.01   01.95.01   1.1.01.01.10   16..10.
```

## SOURCE LISTING

```
C          IN SITU COMBUSTION MODEL
C          THIS PROGRAM PERFORMS A PRELIMINARY ESTIMATE OF OIL RECOVERY
C          AND PERFORMANCE OF IN SITU RECOVERY PROJECT BASED ON EMPIRICAL
C          CORRELATIONS PUBLISHED IN THE U. S. D. O. E. REPORT DOE/ET/12072-2
C          THIS IS NOT A RIGOROUS ANALYSIS AND THE RESULTS SHOULD BE TREATED
C          AS ONLY VERY PRELIMINARY
C
           DIMENSION A(20),E(20),AI(20),XNP(20),ANP(20),AOR(20)
           READ(5,1)(A(I),I=1,20)
           READ(5,1)(B(I),I=1,20)
    1 FORMAT(20A4)
           WRITE(6,2)
    2 FORMAT(//,3X,'PRELIMINARY ESTIMATE OF IN SITU COMBUSTION RWCOVERY'
      *,/,3X,'USES EMPIRICAL CORRELATIONS BASED ON DOE/ET/12072-2 REPORT'
      *)
           WRITE(6,3)(A(I),I=1,20)
           WRITE(6,3)(B(I),I=1,20)
    3 FORMAT(//,20X,20A4)
           READ(5,4)D,H,PERM,POR,SO,BOI
           READ(5,4)SORW,SPACE,AIR,EO2,VIS,YR
    4 FORMAT(6F12.3)
           F=-0.12+0.00262*H+0.000114*PERM+2.23*SO+0.000242*PERM*H/VIS-0.0001
      *89*D-0.0000652*VIS
           XNB=132.*F*SPACE*H
           XN=7758.*SPACE*H*POR*SO/BOI
           CC=0.427*SORW-0.00135*H+2.196*(1./VIS)**0.25
           XX=EO2*POR*SORW/(XN*(1.-POR))
           WRITE(6,5)
           X=AIR*YR*XX
           C=CC*X
           Y=0.47*(1.-EXP(-1.2*C))
           IF(Y.LT.0.)Y=0.
           IYR=YR
           XNP(IYR)=XN*Y-XNB
           SUM=0.
           IF(XNP(IYR).LT.0.)XNP(IYR)=0.
           DO 100 I=1,IYR
           J=I
           SUM=SUM+(AIR*.001/(SPACE*H))
           REC=-0.08667+.18571*SUM
           IF(REC.LT.0.)REC=0.
           IF(REC.GT.1.)REC=1.
           XNP(I)=XNP(IYR)*REC

           IF(I.EQ.1)ANP(I)=XNP(I)
           IF(I.GT.1)ANP(I)=XNP(I)-XNP(I-1)
           IF(ANP(I).GT.0.)AOR(I)=AIR/ANP(I)
           IF(ANP(I).LE.0.)AOR(I)=0.
           IF(AOR(I).GT.40.)GO TO 110
           WRITE(6,6)I,ANP(I),AIR,AOR(I)
    6 FORMAT(1X,I8,6X,F14.0,6X,F13.0,7X,F13.3)
  100 CONTINUE
           GO TO 120
  110 CONTINUE
           J=J-1
  120 CONTINUE
           REC=(XNP(J)/XN)*100.
           CAI=AIR*FLOAT(J)
           CAOR=CAI/XNP(J)
           WRITE(6,7)CAOR
    7 FORMAT(///,3X,'CUM AIR-OIL RATIO',5X,F12.1,1X,'MCF/BBL')
           WRITE(6,8)XNP(J)
    8 FORMAT(/,3X,'OIL RECOVERY',5X,F12.0,1X,'BBL')
           WRITE(6,9)REC
    9 FORMAT(/,3X,'RECOVERY EFFICIENCY',F6.2,1X,'%')
           WRITE(6,11)J
   11 FORMAT(/,3X,'TIME OF INJECTION',3X,I3,1X,'YEARS')
           WRITE(6,12)AIR
   12 FORMAT(/,3X,'ANNUAL AIR INJECTION RATE',3X,F12.0,1X,'MCF/YR')
           WRITE(6,13)CAI
   13 FORMAT(/,3X,'CUM AIR INJECTION',3X,F12.0,1X,'MCF')
    5 FORMAT(//,5X,'YEAR',6X,'OIL PRODUCTIUN',6X,'AIR INJECTION',7X,'AIR
      * OIL RATIO',/60X,/,21X,'BBL',17X,'MCF',16X,'MCF/BBL')
           STOP
           END
//GO.SYSIN DD *
SUSMITA FIELD        MADHUMITA RESERVOIR       NINA SAND
HAMILTON COUNTY           PLAINSBORO      P35        PROG 19
    1035.            19.        500.           0.340        0.718        1.0
      0.693          10.        190000.        0.950        110.0        6.
/*
```

## OUTPUT LISTING

```
PRELIMINARY ESTIMATE CF IN SITU COMBUSTICN RWCOVERY
USES EMPIRICAL CORRELATIONS BASED CN DOE/ET/12072-2 REPORT

              SUSMITA FIELD     MADHUMITA RESERVOIR      NINA SAND

              HAMILTON COUNTY          PLAINSBORC      P35        PROG 19

   YEAR      OIL PRODUCTION       AIR INJECTION      AIR OIL RATIO

               BBL                  MCF                MCF/EBL
     1         8326.              190000.              22.820
     2        15612.              190000.              12.170
     3        15612.              190000.              12.170
     4        15612.              190000.              12.170
     5        15612.              190000.              12.170
     6        13293.              190000.              14.293


CUM AIR-OIL RATIO           13.6 MCF/BBL

CIL RECOVERY            84067. BBL

RECOVERY EFFICIENCY 23.36 %

TIME CF INJECTION       6 YEARS

ANNUAL AIR INJECTION FATE          190000. MCF/YR

CUM AIR INJECTION       1140000. MCF
```

## BIBLIOGRAPHY

U.S. Department of Energy: *Economics of Enhanced Oil Recovery*, Report DOE/ET/ 12072-2 (1981) 16–17.

## PURPOSE

The program provides a preliminary estimate of $CO_2$ flood performance based upon empirical correlations published in U.S. DOE report DOE/ET/12072-2. This is not a rigorous solution; the results are preliminary and should be used with caution.

This work assumes that no response to the $CO_2$ flood will be noticed until 20% hydrocarbon pore volume of $CO_2$ has been injected. Performance calculation is terminated in three years after termination of $CO_2$ injection. $CO_2$ injection is terminated when the incremental $CO_2$-injection-to-oil-recovery ratio becomes greater than 30 MCF/bbl.

## METHOD

$$\text{Oil rec., bbl} = E_V \times P_V \times \left( \frac{S_{orw}}{B_o} - \frac{S_{CO_2}}{B_{CO_2}} \right)(1 - e^{-5.4(V_{CO_2} - 0.2)})$$

where

$E_v$ = Water flood sweep, fraction;
$P_v$ = Pore volume, bbl;
$S_{orw}$ = Water flood residual oil saturation, fraction;
$S_{CO_2}$ = $CO_2$ flood residual oil saturation, fraction;
$B_o$ = Oil formation volume factor after waterflood, bbl/STB;
$B_{CO_2}$ = Oil formation volume factor after $CO_2$ flood, bbl/STB;
$V_{CO_2}$ = Injected volume of $CO_2$, fraction or multiple of hydrocarbon pore volume.

## PROGRAM DESCRIPTION

The program consists of a simple FORTRAN main program. Users provide annual $CO_2$ injection volumes as a fraction of hydrocarbon pore volume. However, if the incremental $CO_2$ injection volume exceeds 30 MCF per barrel of incremental oil recovery, the program automatically terminates further $CO_2$ injection. However, performance calculations are carried out for another three-year period since $CO_2$ is still in the reservoir and energy is still available because of continued water injection after the $CO_2$ injection is terminated. For waterflood sweeps of 60% or less, the $CO_2$ flood sweep is assumed to be the same as the waterflood sweep. However, if the waterflood sweep is greater than 60%, a correction factor of 0.75 is utilized for the $CO_2$ flood sweep to account for viscous fingering of $CO_2$ due to an adverse mobility ratio.

### Input Format

| Card | Format | Content |
|------|--------|---------|
| 1 | 20A4 | A(I) |
| 2 | 20A4 | B(I) |
| 3 | F5.3, F10.0, 2F5.3, 2F5.2, 2F5.0 | EV, PV, SORW, $SCO_2$, BO, $BCO_2$, PR, TR |
| 4 | I2 | N |
| 5 | 20F4.3 | $VCO_2$(I) |

### Description of Input Parameters

$A(I)$ = Uses up to 80 alphanumeric characters for project identification;

$B(I)$ = Uses up to 80 alphanumeric characters for project identification;

$EV$ = Waterflood volumetric sweep efficiency, fraction;

$PV$ = Reservoir pore volume, bbl;

$SORW$ = Waterflood residual oil saturation, fraction;

$SCO_2$ = $CO_2$ flood residual oil saturation, fraction;

$BO$ = Oil formation volume factor after waterflood, bbl/STB;

$BCO_2$ = Oil formation volume factor after $CO_2$ flood, bbl/STB;

$PR$ = Reservoir pressure ($CO_2$ flood), psi;

$TR$ = Reservoir temperature, °F;

$N$ = Number of years performance is desired (not to exceed 20 years);

$VCO_2(I)$ = Annual $CO_2$ injection volume, fraction or multiple of hydrocarbon pore volume.

**EXAMPLE PROBLEM**

Compute $CO_2$ flood performance for a given reservoir for 10% hydrocarbon pore volume of $CO_2$ injection per year for 12 years. The following additional information is available:

Waterflood volumetric sweep = 54%,
Reservoir pore volume = 1490000 bbl,
Waterflood residual oil saturation = 35%,
$CO_2$ flood residual oil saturation = 31.5%,
Oil formation volume factor after waterflood = 1.08 bbl/STB,
Oil formation volume factor after $CO_2$ flood = 1.62 bbl/STB,
Reservoir pressure = 2500 psi,
Reservoir temperature = 100° F.

**INPUT LISTING**

FORTRAN CODING F

WW/CC

NAME
DATE

PROGRAM
ROUTINE

STATEMENT NUMBER

| 1 2 3 4 5 | 6 | 7 8 9 10 11 12 13 14 15 16 17 18 19 20 21 22 23 24 25 26 27 28 29 30 31 32 33 34 35 36 37 38 39 40 41 42 43 44 45 46 47 48 49 50 |

O = ZERO         I = ONE
Ø = ALPHA O      1 = ALPHA 1

CARD
1. PRØJECT | ABAMIS CHANDRA | P.4
2. GØMIA | 616.1DIH BIHAR
3. .59 | .1490.00...35 .315 .1.018 .1.6.25.0.0. .100.
4. 12
5. .1

**SOURCE LISTING**

```
C     CARBON DIOXIDE FLOOD MODEL
C     THIS PROGRAM PERFORMS A PRELIMINARY ESTIMATE OF OIL RECOVERY
C     AND PERFORMANCE OF CARBON DIOXIDE FLOOD PROJECT BASED ON EMPRIICAL
C     CORRELATIONS PUBLISHED IN THE U.S.D.O.E. REPORT DOE/ET/12072-2
C     THIS IS NOT A RIGOROUS ANALYSIS AND THE RESULTS ARE ONLY FIRST
C     ORDER APPROXIMATICNS AND SHOULD BE TREATED AS SUCH.
C
      DIMENSION A(20),B(20),VCO2(20),CREC(20),ANREC(20)
C
C     EV=WATERFLOOD SWEEP EFFICIENCY, FRACTION
C     PV=PORE VOLUME, BBL
C     SCRW=WATERFLOOD RESIDUAL OIL SATURATION, FRACTION
C     BO=OIL FORMATION VOLUME FACTOR AFTER WATERFLOOD, RES BBL/STB
C     BCO2=INJECTED VOLUME OF CO2, FRACTION OF PORE VOLUME
C     TR=RESERVOIR PRESSURE, PSIA
C     TR=RESERVOIR TEMPERATURE, DEG F
C
      READ(5,1)(A(I),I=1,20)
      READ(5,1)(B(I),I=1,20)
    1 FORMAT(20A4)
      WRITE(6,2)
    2 FORMAT(//,3X,'PRILIMINARY ESTIMATE OF CO2 FLOOD PERFORMANCE')
      WRITE(6,3)
    3 FORMAT(/,3X,'BASED ON EMPIRICAL CORRELATIONS PUBLISHED IN THE DOE
     *REPORT DOE/ET/12072-2')
      WRITE(6,4)(A(I),I=1,20)
      WRITE(6,4)(B(I),I=1,20)
    4 FORMAT(//,20X,20A4)
      WRITE(6,11)
   11 FORMAT(//,1X,'YEAR',4X,'OIL PRODN',6X,'CO2/OIL',3X,'CO2 INJ')
      WRITE(6,12)
   12 FORMAT(12X,'BBL',10X,'RATIO',6X,'MCF')
      WRITE(6,13)
   13 FORMAT(25X,'MCF/BBL')
      READ(5,5)EV,PV,SORW,SCO2,BO,BCO2,PR,TR
    5 FORMAT(F5.3,F10.0,2F5.3,2F5.2,2F5.0)
      READ(5,6)N
    6 FORMAT(I2)
      READ(5,7)(VCO2(I),I=1,20)
    7 FORMAT(20F4.3)
  100 CONTINUE
      X=EV*PV*(SORW/BO-SCO2/BCO2)
      IF(EV.GT.0.6)X=0.75*X
      SUM=0.0
      DO 200 I=1,N
      II=I
      SUM=SUM+VCO2(I)
      IF(SUM.LE.0.21)GO TO 150
      XX=(1.-EXP(-5.4*(SUM-0.2)))
      CREC(I)=X*XX
      ANREC(I)=CREC(I)-CREC(I-1)
      IF(ANREC(I).LE.0.01)GO TO 150
      CO2INJ=0.1986*VCO2(I)*PV*(PR/(TR+460.))
      IF(ANREC(I).GT.0.0)CO2OR=CO2INJ/ANREC(I)
      GO TO 160
    9 FORMAT(I5,4X,F9.0,6X,F7.2,3X,F10.2)
  150 CONTINUE
      CO2INJ=0.1986*VCO2(I)*PV*(PR/(TR+460.))
      ANREC(I)=0.
      CREC(I)=0.
      CO2OR=0.
  160 CONTINUE
      WRITE(6,9)I,ANREC(I),CO2OR,CO2INJ
      IF(CO2OR.GT.30.)GO TO 210
  200 CONTINUE
  210 CONTINUE
      NN=II+3
      III=II
      DO 215 J=III,NN
      VCO2(J)=VCO2(II)
      SUM=SUM+VCO2(J)
      XX=(1.-EXP(-5.4*(SUM-0.2)))
      CREC(J)=X*XX
      ANREC(J)=CREC(J)-CREC(J-1)
      CO2OR=0.0
      CO2INJ=0.
      WRITE(6,9)J,ANREC(J),CO2OR,CO2INJ
  215 CONTINUE
      TREC=(CREC(NN)/(PV*SORW/BO))*100.
      WRITE(6,15)CREC(NN)
```

```
   15 FORMAT(//,3X,' CO2 RECOVERY',3X,F10.0,1X,'BBL')
      WRITE(6,10)TREC
   10 FORMAT(//,3X,'ULTIMATE CO2 FLOOD RECOVERY',1X,F10.1,' X')
      STOP
      END
//GO.SYSIN DD *
      PROJECT   ABANIS CHANDRA               P 41
      GOMIA     GRIDIH            BIHAR
   .540   1490000.  .350  .315 1.08 1.622500. 100.
12
.1   .1   .1   .1   .1   .1   .1   .1   .1   .1   .1
/*
```

## OUTPUT LISTING

```
PRELIMINARY ESTIMATE CF CO2 FLOOD PERFORMANCE

BASED ON EMPIRICAL CORRELATIONS PUBLISHED IN THE DOE REPORT DOE/ET/12072-2


                    PROJECT   ABANIS CHANDRA        P 41


                    GOMIA     GRIDIH        BIHAR


YEAR      OIL PRODN     CO2/OIL     CO2 INJ
             BBL        RATIO         MCF
                        MCF/BBL
  1           0.          0.0      132104.44
  2           0.          0.0      132104.44
  3        43519.         3.04     132104.44
  4        25361.         5.21     132104.44
  5        14779.         8.94     132104.44
  6         8612.        15.34     132104.44
  7         5019.        26.32     132104.44
  8         2925.        45.17     132104.44
  8         4629.         0.0          0.0
  9          993.         0.0          0.0
 10          579.         0.0          0.0
 11          337.         0.0          0.0


   CO2 RECOVERY        103829. BBL

ULTIMATE CO2 FLOOD RECOVERY        21.5 %
```

## BIBLIOGRAPHY

U.S. Department of Energy: *Economics of Enhanced Oil Recovery*, Report DOE/ET/12072-2 (1981) 17–18.

# 15    POLYMER FLOOD PERFORMANCE

## PURPOSE

This program provides a preliminary estimate of polymer flood performance and recovery based on empirical correlations published in U.S. DOE Report DOE/ET/12072-2. The program does not, however, provide a rigorous solution. The results obtained are preliminary in nature and should be treated as such.

Although the theoretical basis for increased oil recovery by polymer flood is sound, actual field performance is usually lower than that predicted by analytical solutions or laboratory experiment. Empirical data suggest that the addition of polymer improves sweep by 1% for every 10% that the primary/secondary sweep is less than 100%; that is, the recoveries are inversely dependent on the effectiveness of the waterflood sweep.

The estimate provided by this program will indicate relatively good or improved flood performance when waterflood results are poor. Conversely, however, the estimate will show relatively insignificant performance improvement when the waterflood performance has been good.

## METHOD

$$N_p = \left[ \left( 0.1 - \frac{E_v}{10} \right) \left( \frac{S_{oi} - S_{orp}}{S_{oi}} \right) + E_v \left( \frac{S_{orw} - S_{orp}}{S_{oi}} \right) \right] \times \frac{B_{oi}}{B_o} \times N$$

where

$N_p$ = Incremental recovery due to polymer flooding, bbl;
$E_v$ = Waterflood sweep, percent;
$N$ = Original oil in place, bbl;
$B_{oi}$ = Initial oil formation volume factor, bbl/STB;
$B_o$ = Oil formation volume factor after primary and secondary recovery;

147

$S_{oi}$ = Initial oil saturation;
$S_{orw}$ = Residual oil saturation in water-swept region;
$S_{orp}$ = Residual oil saturation to polymer flooding.

For scheduling the production, an exponentially declining pseudo-sweep as a function of the cumulative pore volume of polymer-augmented water injection has been derived. These pseudo-sweep values with a lower limit equal to the waterflood sweep are used in conjunction with the equation to provide cumulative production.

## PROGRAM DESCRIPTION
This program consists of a simple FORTRAN main program and computes performance for a maximum of 20 years based on the user-supplied annual volumes of polymer-augmented water injection (fraction of reservoir pore volume).

### Input Format

| Card | Format | Content |
|---|---|---|
| 1 | 20A4 | A(I) |
| 2 | 20A4 | B(I) |
| 3 | F5.0, F5.1, F5.2, F5.1, F5.2 | SPACE, H, POR, SOI, BOI |
| 4 | 2F5.1, F5.2, F5.1, F5.2 | SORW, SORP, BO, EV, POLCON |
| 5–N | 20F4.2 | PV(I) |

### Description of Input Parameters

$\quad\quad$ A(I) = Uses up to 80 alphanumeric characters for project identification;
$\quad\quad$ B(I) = Uses up to 80 alphanumeric characters for project identification;
$\quad$ SPACE = Pattern area, acre;
$\quad\quad\quad$ H = Net pay thickness, ft;
$\quad\quad$ POR = Average reservoir porosity, percent;
$\quad\quad$ SOI = Initial oil saturation, percent;
$\quad\quad$ BOI = Initial oil formation volume factor, bbl/STB;
$\quad$ SORW = Residual oil saturation in water-swept region, percent;
$\quad$ SORP = Residual oil saturation in polymer-swept region, percent;
$\quad\quad$ BO = Oil formation volume factor at secondary recovery, bbl/STB;
$\quad\quad$ EV = Waterflood sweep (anticipated and/or actual), percent;
POLCON = Polymer concentration, lb/bbl;
$\quad$ PV(I) = Fraction pore volume polymer solution injection.

**EXAMPLE PROBLEM**

Determine polymer flood performance for a reservoir with the following characteristics. The polymer solution injection is continued for 6 years at an annual rate of 10% of pore volume.

Depth = 3380 ft,
Thickness = 27 ft,
Initial oil saturation = 62.8%,
Residual oil saturation, after waterflood = 39.7%,
Residual oil saturation after polymer flood = 38.9%,
Average porosity = 29%,
Initial formation volume factor = 1.05 bbl/STB,
Oil formation volume factor at the end of waterflood = 1.05 bbl/STB,
Waterflood sweep = 40%.

**INPUT LISTING**

FORTRAN CODING F

UW/CC

NAME
DATE

0 = ZERO
Ø = ALPHA O

1 = ONE
1 = ALPHA 1

PROGRAM
ROUTINE

| CARD | STATEMENT NUMBER 1,2,3,4,5 | 6 | 7,8,9,10,11,12,13,14,15,16,17,18,19,20,21,22,23,24,25,26,27,28,29,30,31,32,33,34,35,36,37,38,39,40,41,42,43,44,45,46,47,48,49,50 |
|---|---|---|---|
| 1. | N,I,N,A | | E,H,Ø,S,H |  |  |  |  |  |  |  D,N,I,P, ,E,H,Ø,S,H, , , , , ,B,L,A,N,I, ,E,H,Ø,S,H |
| 2. | S,A,U,G,A,S | | | | | C,A,L,I,F,Ø,R,N,I,A |
| 3. | 4,0. | | 2,7. ,2,9. ,6,2. ,8, ,1. ,0,5 |
| 4. | 3,9. ,7 | | 3,8. ,9, ,1. ,0,5, ,4,0. ,2. ,8,8 |
| 5. | 1,0. | | 1,0. ,1,0. ,1,0. ,1,0. ,1,0. ,1,0. |

## SOURCE LISTING

```
C          THIS PROGRAM PERFCRMS A PRELIMINARY ESTIMATE OF POLYMER-FLOOD
C          PERFORMANCE BASED ON EMPIRICAL CORRELATION PUBLISHED IN THE
C          U.S.D.O.E. REPORT DOE/ET/12072-2.  THIS IS NOT A RIGOROUS ANALYTICA
C          PROCEDURE AND THE RESULTS SHCULD BE TREATED AS ONLY PRELIMINARY
C
           DIMENSION A(20),B(20),PV(20),CUMREC(20),ANREC(20),POLOR(20)
          *,POLVOL(20)
           WRITE(6,1)
         1 FCRMAT(//,3X,'PRELIMINARY ESTIMATE OF POLYMERFLOOD PERFORMANCE')
           WRITE(6,2)
         2 FORMAT(3X,'THIS ESTIMATE SHOULD BE  TREATED AS FIRST ORDER APPROXIM
          *ATION ONLY')
           READ(5,3)(A(I),I=1,20)
           READ(5,3)(B(I),I=1,20)
         3 FORMAT(20A4)
           WRITE(6,4)(A(I),I=1,20)
           WRITE(6,4)(B(I),I=1,20)
         4 FORMAT(/,20X,20A4)
C          EV=ESTIMATED WATER-FLOOD SWEEP EFFICIENCY, %
C          SOI=INITIAL OIL SATURATICN, %
C          SORW=RESIDUAL OIL SATURATION IN WATER SWEPT REGION, %
C          SORP=RESIDUAL CIL SATURATION IN POLYMER SWEPT REGION, %
C          BOI=INITIAL OIL FVF, BBL/STB
C          SPACE=PATTERN SPACING, AC
C          H=AVG. NET PAY IN THE PATTEFN AREA, FT
C          POR=AVG. POROSITY, PER CENT
C          PV=PORE VOLUME PCLYMER INJECTED, FRACTION
C          POLCON=POLUMER CONCENTRATICN, LB/BBL
C          POLOR=POLYMER/CIL RATIO, LB/BBL OF OIL PRODUCED
C
           READ(5,6)SPACE,H,POR,SOI,BOI
           READ(5,7)SORW,SCRP,BO,EV,POLCON
         6 FCRMAT(F5.0,F5.1,F5.2,F5.1,F5.2)
         7 FORMAT(2F5.1,F5.2,F5.1,F5.2)
           OOIP=0.7758*SPACE*H*POR*SCI/BOI
           READ(5,8)(PV(I),I=1,20)
         8 FORMAT(20F4.2)
           SUM1=0.0
           DO 120 I=1,20
           PCLVOL(I)=0.01*PV(I)*7758.*SPACE*H*POR*(.01*EV+(0.1-0.001*EV))
           SUM1=SUM1+POLVCL(I)
       120 CONTINUE
           POLUSD=SUM1/PCLCCN
           OILREC=((0.1-0.001*EV)*(SCI-SORP)/SOI+.01*EV*(SORW-SCRP)/SOI)*
          *BCI*OOIP/BO
           POLOR(1)=POLUSD/OILREC
           WRITE(6,15)POLOR(1)
        15 FORMAT(//,3X,'POLYMER USE OIL RATIO',3X,F5.2,1X,'LB/BBL OF ',
          *'OIL RECOVERY')
           SUM=0.
           SUM1=0.0
           J=0
           DO 130 I=1,20
           IJ=I
           IF(PV(I).EQ.0.)J=J+1
           IF(J.GE.4) GO TO 135
       130 CONTINUE
       135 II=IJ
           DO 100 I=1,II
           IF(PV(I).EQ.0.)PV(I)=.1
           SUM=SUM+PV(I)
           EV1=EXP(-1.6109*SUM+5.298)
           EV1=AMAX1(EV1,EV)
           CUMREC(I)=((0.1-0.001*EV1)*(SOI-SORP)/SOI+.01*EV1*(SORW-SORP)/SOI)
          **BOI*OOIP/BO
           IF(CUMREC(I).LE.0.0)CUMREC(I)=0.
           IF(I.EQ.1)ANREC(I)=CUMREC(I)
           IF(I.GT.1)ANREC(I)=CUMREC(I)-CUMREC(I-1)
           IF(ANREC(I).GT.0.)POLOR(I)=POLOR(I)/(POLCON*ANREC(I))
           IF(ANREC(I).LE.0.)PCLOR(I)=0.
           SUM1=SUM1+ANREC(I)
           OLREC1=(SUM1/OCIP)*100.
       100 CONTINUE
           WRITE(6,11)
        11 FORMAT(///,5X,'YEAR',6X,'CIL PRODUCTION',10X,'POLYMER INJ')
           WRITE(6,12)
        12 FORMAT(20X,'BBL',21X,'BBL')
           DO 150 I=1,IJ
           WRITE(6,13)I,ANREC(I),POLVOL(I)
        13 FORMAT(5X,I5,5X,F15.0,9X,F11.0,5X,F13.2)
```

```
   150 CONTINUE
       WRITE(6,20)OO IP
       WRITE(6,21)SUM1
       WRITE(6,22)OLREC1
    20 FORMAT(/,3X,'OIL-IN-PLACE',3X,F15.0,' BBL')
    21 FORMAT(/,3X,'OIL RECOVERY',3X,F15.0,' BBL')
    22 FORMAT(/,3X,'OIL RECOVERY',3X,F15.5,' X')
       STOP
       END
//GO.SYSIN DD *
       NINA GHOSH          DWIP GHOSH      BANI GHOSH
       SAUGAS             CALIFORNIA
    40. 27.029.00 62.8 1.05
   39.7 38.9 1.05 40.0 2.88
    .10 .10 .10 .10 .10 .10 .0  .0  .0  .0  .0  .0  .0  .0  .0  .0  .0  .0
/*
```

## OUTPUT LISTING

```
PRELIMINARY ESTIMATE CF POLYMERFLOOD PERFORMANCE
THIS ESTIMATE SHOULD BE TREATED AS FIRST ORDER APPROXIMATICN ONLY

                    NINA GHCSH        DWIP GHOSH     BANI GHOSH

                    SAUGAS            CALIFORNIA

POLYMER USE OIL RATIO    5.74 LB/BBL OF OIL RECOVERY


   YEAR        OIL PRCDUCTION            POLYME R INJ
                    BBL                      BBL
    1                 0.               111771.
    2              2004.               111771.
    3              7931.               111771.
    4              6751.               111771.
    5              5746.               111771.
    6              4891.               111771.
    7              4164.                    0.
    8              3544.                    0.
    9              3017.                    0.
   10              2542.                    0.

OIL-IN-PLACE         1453254. BBL

OIL RECOVERY           40589. BBL

OIL RECCVERY          2.79299 X
```

## BIBLIOGRAPHY

U.S. Department of Energy: *Economics of Enhanced Oil Recovery*, Report DOE/ET/12072-2 (1981) 19–20.

# IV RESERVOIR ENGINEERING FOR NATURAL GAS

PURPOSE

The program provides values of the natural gas compressibility factor $(Z)$ and gas viscosity $(\mu_g)$ at desired temperatures and pressures. Either gas gravity or gas composition should be known. Corrections for impurities such as carbon dioxide, nitrogen, and hydrogen sulfide are applied (if the mol fraction composition of the impurities is known).

For $Z$-factor calculation, the Hall-Yarborough procedure is utilized. The most widely used Standing-Katz $Z$-factor chart terminates at reduced pressure of 24. However, the Hall-Yarborough procedure is claimed to be valid at even higher pressure.

The only accurate way to obtain the viscosity of a gas is to determine it experimentally; however, experimental determination is difficult. Thus, practicing reservoir engineers usually rely on viscosity correlations developed from the equation of state and the theorem of corresponding states. There are various different forms of such correlations. This program uses one such correlation that takes into account the presence of low concentrations of carbon dioxide, nitrogen, and hydrogen sulfide. The other limitations of reduced pressure and reduced temperature are $P_r \le 20$ and $1.2 \le T_r \le 3$.

**METHOD**

$$Z = (1 + Y + Y^2 - Y^3)/(1 - Y)^3 - (14.76t - 9.76t^2 + 4.58t^3)Y$$
$$+ (90.7t - 242.2t^2 + 42.4t^3)Y^{(1.18 + 2.82t)},$$

$$\frac{bp_c}{RT_c} = 0.245 \exp[-1.2(1 - t)^2],$$

$$Y = \frac{bp}{4},$$

where

$t$ = Reciprocal reduced temperature, °R;
$p_c$ = Critical pressure;
$T_c$ = Critical temperature;
$R$ = Universal gas constant;
$\rho$ = Density;
$b$ = Van der Waal's–type covolume;
$Z$ = Compressibility factor.

## PROGRAM DESCRIPTION

The FORTRAN program consists of a main program and two subprograms. The main program is the controlling program and determines critical pressure and critical temperature values from either the gas composition or the gas gravity values provided.

Subprogram COMFAC determines the compressibility factor values, while subprogram VISCO computes the viscosity values. The program generates a set of $Z$-factors and gas viscosity values for the gas at any given reservoir temperature and a set of pressure values.

### Input Format

| Card | Format | Content |
|---|---|---|
| 1 | 20A4 | A(I) |
| 2 | 2I2, F4.0, F4.1 | N, IOPT, TR, BO |
| 3 | 13F5.4 | C1, C2, C3, XIC4, XNC4, |
| (Required if | | XIC5, XNC5, XNC6, |
| IOPT = 2 | | XNC7, C7P, XN2, |
| in card 2) | | CO2, H2S |
| 4 | 4F5.4 | GG, XN2, CO2, H2S |
| (Required if | | |
| IOPT = 2 | | |
| in card 2) | | |
| 5–N | F5.0 | PR |

### Description of Input Parameters

A(I) = Uses up to 80 alphanumeric characters for project identification and description, such as
Example Problem 2   Morgantown   West Virginia;
N = Number of pressure points at which physical properties are to be evaluated;

IOPT = Option Flag: IOPT = 2, gas gravity and molecular fraction impurities are provided; IOPT ≠ 2, gas composition is provided;

  TR = Reservoir temperature or temperature at which Z-factor values are to be computed;

  BO = Barometric pressure, psia;

  C1 = Mol fraction of methane;

  C2 = Mol fraction of ethane;

  C3 = Mol fraction of propane;

XIC4 = Mol fraction of isobutane;

XNC4 = Mol fraction of N-butane;

XIC5 = Mol fraction of I-pentane;

XNC5 = Mol fraction of N-pentane;

XNC6 = Mol fraction of N-hexane;

XNC7 = Mol fraction of N-heptane;

  C7P = Mol fraction of $C_7$ plus;

  XN2 = Mol fraction of nitrogen;

  CO2 = Mol fraction of carbon dioxide;

  H2S = Mol fraction of hydrogen sulfide;

  GG = Gas gravity (air = 1);

  PR = Pressures at which properties are to be computed, should correspond with $N$ (card 2).

### EXAMPLE PROBLEM

Determine gas properties (compressibility factor $Z$, gas viscosity) at pressures of 3700, 3200, 2700, 2200, 1700, 1200, 700, 400, and 100 and at the reservoir temperature for a gas with the composition shown in table 16.1:

  Reservoir temperature = 230° F;
  Barometric pressure = 14.7 psia.

**Table 16.1 Gas Composition**

| Component | Mol Fraction |
|---|---|
| C1 | 0.8794 |
| C2 | 0.0555 |
| C3 | 0.0252 |
| IC4 | 0.0064 |
| NC4 | 0.0068 |
| IC5 | 0.0045 |
| NC5 | 0.0032 |
| NC6 | 0.0027 |
| NC7 | 0.0000 |
| C7 + | 0.0163 |
| $N_2$ | 0.0000 |
| $CO_2$ | 0.0000 |
| $H_2S$ | 0.0000 |

**INPUT LISTING**

# FORTRAN CODING FORM

UU/CC   NAME ___
        DATE ___

0 = ZERO   1 = ONE   2 = TWO
Ø = ALPHA O   I = ALPHA I   Z = ALPHA Z

| STATEMENT NUMBER | | 6 | 7,8,9... (statement text) |
|---|---|---|---|
| CARD 1 | | | EXAMPLE PROBLEM 2   MØRGAINTØWN   WGISTVIRGIWIIA |
| 2. | 91123 | | 0..14.7 |
| 3. | 8794. | | .0555..0252..00064..00168..00145..000312..010217..010000..010163.10000.010010.1010 |
| 5-1 | 37000. | | |
| 5-2 | 32.00. | | |
| 5-3 | 27.00. | | |
| 5-4 | 22.00. | | |
| 5-5 | 17.00. | | |
| 5-6 | 12.00. | | |
| 5-7 | 700. | | |
| 5-8 | 400. | | |
| 5-9 | 100. | | |

## SOURCE LISTING

```
C       THIS PROGRAM IS DESIGNED TO COMPUTE NATURAL GAS PROPERTIES FOR A
C       GIVEN GAS STREAM AT REQUIRED PRESSURE AND TEMPERATURE.  EITHER
C       THE GRAVITY OR THE COMPOSITION OF THE GAS MUST BE KNOWN.
C
C       IOPT=OPTION FLAG FOR COMPOSITION OR GRAVITY
C       IOPT=1; COMPOSITION IS GIVEN
C       IOPT=2; COMPOSITION IS NOT AVAILABLE. GAS GRAVITY IS GIVEN
C       N=NUMBER OF PRESSURES AT WHICH PROPERTIES ARE TO BE CALCULATED
C       TR=TEMPERATURES AT WHICH GAS PROPERTIES ARE TO BE CALCULATED
        DIMENSION A(20)
        READ(5,61)(A(I),I=1,20)
   61   FORMAT(20A4)
        READ(5,1)N,IOPT,TR,BO
    1   FORMAT(2I2,F4.0,F4.1)
        IF(IOPT.EQ.2) GO TO 20

C       READ MOL. FRACTION GAS COMPOSITION
C
        READ(5,2)C1,C2,C3,XIC4,XNC4,XIC5,XNC5,XNC6,XNC7,C7P,XN2,CO2,H2S
    2   FORMAT(13F5.4)
        XPC=667.8*C1+707.8*C2+616.3*C3+529.1*XIC4+550.7*XNC4+490.4*XIC5+
       *488.6*XNC5+436.9*XNC6+100.205*XNC7+114.23*C7P+493.6*XN2+1071.*CO2
       *+1306.*H2S
        XTC=-116.63*C1+90.09*C2+206.01*C3+274.98*XIC4+305.65*XNC4+369.10*
       *XIC5+385.7*XNC5+453.7*XNC6+512.8*XNC7+564.22*C7P-232.4*XN2+87.9*C
       *O2+212.7*H2S
        TC=XTC+460.
        PC=XPC
        XMOL=16.043*C1+30.07*C2+44.097*C3+58.124*(XIC4+XNC4)+72.151*(XIC5+
       *XNC5)+86.178*XNC6+100.205*XNC7+114.232*C7P+28.013*XN2+44.01*CO2+34
       *.076*H2S
        CGG=0.03453*XMOL
        GG=CGG
        GO TO 40
   20   CONTINUE
        READ(5,3)GG,XN2,CO2,H2S
    3   FORMAT(4F5.4)
        PC=676.2366+6.9868*GG-25.8273*GG**2+432.9*CO2-167.3*XN2+654.*H2S
        TC=142.7712+380.9671*GG-36.4051*GG**2-167.3*CO2-279.9*XN2+127.*H2S
   40   CONTINUE
        WRITE(6,62)(A(I),I=1,20)
   62   FORMAT(///,3X,20A4)
        WRITE(6,100)
  100   FORMAT(//,5X,'PRESSURE',3X,'COMP.FAC',3X,'VISCO.CP')
        DO 50 I=1,N
        READ(5,7)PR
    7   FORMAT(F5.0)
        PR=PR+BO
        CALL COMFAC(PR,TR,PC,TC,Z)
        CALL VISCO(PR,TR,PC,TC,GG,CO2,XN2,H2S,VIS)
        WRITE(6,8) PR,Z,VIS
    8   FORMAT(/,4X,F8.0,4X,F8.5,3X,F8.5)
   50   CONTINUE
   60   CONTINUE
        STOP
        END
        SUBROUTINE COMFAC(T,S,PC,TC,Z)
        PR=T
        X=S+460.
    5   RRT=TC/X
    6   RP=PR/PC
    7   A=0.06125*RRT*EXP(-1.2*(1.-RRT)**2)
    8   B=RRT*(14.76-9.76*RRT+4.58*RRT**2)
    9   C=RRT*(90.7-242.2*RRT+42.4*RRT**2)
        D=2.18+2.82*RRT
        Y=0.01
        DO 2 J=1,30
        IF(Y.GT.1.)Y=0.6
   10   F=-A*RP+Y*(1.+(Y*(1.+Y*(1.-Y))))/(1.-Y)**3-B*Y**2+C*Y**D
        IF(ABS(F)-1.0E-6)4,4,3
    3   DFDY=(1.+4.*Y*(1.+Y*(1.-Y))+Y**4)/(1.-Y)**4-2.*B*Y+D*C*Y**(D-1.)
    2   Y=Y-F/DFDY
    4   Z=A*RP/Y
        RETURN
        END
        SUBROUTINE VISCO(PB,TRE,PC,TC,GG,C1,C2,C3,VIS)
        REAL M
        DATA B0,B1,B2,B3,B4,B5,B6,B7,B8/1.112391E-2,1.677266E-5,2.113605E
       *-9,-1.094850E-4,-6.403164E-8,8.993745E-11,4.577352E-7,2.129034E-1
       *0,3.977322E-13/
```

```
      DATA A0,A1,A2,A3,A4,A5,A6,A7,A8,A9/-2.462118,2.970547,-2.862641E-1
     *,8.054205E-3,2.808609,-3.498033,3.603730E-1,-1.044324E-2,-7.93385
     *7E-1,1.396433/
      DATA H0,H1,H2,H3,H4,H5/-1.491449E-1,4.410155E-3,8.393872E-2,-1.86
     *4088E-1,2.033679E-2,-6.095793E-4/
      TR=TRE+460.
      RP=PB/PC
      RT=TR/TC
      IF(RT.GT.3.)GO TO 20
      IF(RT.LT.1.2)GO TO 30



      IF(RP.GT.20.) GO TO 40
      M=28.97*GG
      Y5=GG/(39.8782+64.0537*GG)
      Y6=1.03401E-2-3.4823E-3/GG
      Y7=8.08098E-3-3.99988E-3/GG
      U=TR-460.
      V1=C1*Y5+C2*Y6+C3*Y7
      V2=B0+B1*U+B2*U**2+B3*M+B4*U*M+B5*U**2*M+B6*M**2+B7*U*M**2+B8*U**
     *2*M**2
      V3=((V1+V2)*1.E05+0.5)*1.E-5
      IF(RP.LT.1.)GO TO 50
      V4=A0+A1*RP+A2*RP**2+A3*RP**3+RT*(A4+A5*RP+A6*RP**2+A7*RP**3)
      V5=RT**2*(A8+A9*RP+H0*RP**2+H1*RP**3)
      V6=RT**3*(H2+H3*RP+H4*RP**2+H5*RP**3)
      V7=V4+V5+V6
      V8=EXP(V7)/RT
      U1=((V3*V8)*1.E4+0.5)*1.E-4
      GO TO 65
   20 WRITE(6,21)
   21 FORMAT(//,3X,'PSEUDO REDUCED TEMPERATURE IS GREATER THAN 3.0')
   30 WRITE(6,31)
   31 FORMAT(//,3X,'PSEUDO REDUCED TEMPERATURE IS LESS THAN 1.2')
   40 WRITE(6,41)
   41 FORMAT(//,3X,'PSEUDO REDUCED PRESSURE IS GREATER THAN 20')
      GO TO 60
   50 U1=V3
      GO TO 65
   60 U1=0.
   65 CONTINUE
      VIS=U1
      RETURN
      END
//GO.SYSIN DD *
  EXAMPLE PROBLEM 2   MORGANTOWN   WEST VIRGINIA
  9 1230.14.7
 .8794.0555.0252.0064.0068.0045.0032.0027.0000.0163
3700.
3200.
2700.
2200.
1700.
1200.
 700.
 400.
 100.
/*
```

## OUTPUT LISTING

```
EXAMPLE PROBLEM 2   MORGANTOWN   WEST VIRGINIA

PRESSURE     COMP.FAC     VISCO.CP
   3715.       0.92920      0.02105
   3215.       0.90741      0.01969
   2715.       0.89414      0.01836
   2215.       0.89120      0.01708
   1715.       0.89953      0.01584
   1215.       0.91871      0.01465
    715.       0.94718      0.01352
    415.       0.96785      0.01289
    115.       0.99074      0.01289
```

## BIBLIOGRAPHY

Hall, Kenneth R., and Yarborough, Lyman: "A New Equation of State for Z-Factor Calculations," *Oil and Gas J.* (June 18, 1973) 82–92.

Yarborough, Lyman, and Hall, Kenneth R.: "How to Solve Equation of State for Z-Factor," *Oil and Gas J.* (February 18, 1974) 86–88.

# 17    DETERMINATION OF GAS IN PLACE BY THE MATERIAL BALANCE METHOD FOR A WATER DRIVE RESERVOIR

## PURPOSE

Material balance calculations are performed to determine the original gas-in-place estimate, utilizing actual performance data. The method is based on the Havlena and Odeh procedure. Use is made of the Van Everdingen-Hurst $Q_T$ function values for water influx computation. $Q_T$ versus $T_D$ values that are based on an assumed aquifer to reservoir size are supplied by the user from published literature. It is also assumed that the gas reservoir has radial geometry.

The suggested procedure to check the validity of aquifer size and shape is to prepare a coordinate plot of a material balance variable ($Y$) versus a water influx variable ($X$) that is computed in the program and printed in the output listing. A linear plot will indicate that the assumptions made are valid. If the plot so obtained is not linear, then the aquifer size assumption must be changed and the program rerun. If a trial-and-error selection of the aquifer size does not eventually produce a linear plot, then the reservoir geometry and boundary condition assumptions are not valid. This nonlinearity then implies that the estimated original gas-in-place value obtained by using this program is not reliable.

## METHOD

$$\frac{G_p B_g + W_p}{B_g - B_{gi}} = G + \frac{B\Sigma\Delta PQ_{T_D}}{B_g - B_{gi}},$$

$$\frac{G_p B_g + W_p}{B_g - B_{gi}} = G + \frac{B\sum_{i=1}^{n} \Delta P_{avgi} Q_{T_D}[t(n) - t(i - 1)]}{B_g - B_{gi}}.$$

Let

$$\frac{G_p B_g + W_p}{B_g - B_{gi}} = Y, \text{ called the material balance variable,}$$

and

$$\frac{\sum_{i=1}^{n} \Delta P_{avgi} Q_{T_D}[t(n) - t(i - 1)]}{B_g - B_{gi}} = X, \text{ called the water influx variable.}$$

Then

$$Y = G + BX. \tag{17.1}$$

If correct assumptions of reservoir geometry, aquifer size, and dimensionless time parameters are made, then a linear plot will result as indicated by equation (17.1). The ordinate intercept ($G$) is the original gas in place, and the slope of the line (B) is the water influx constant.

$$T_D = \frac{2.31045k}{\phi \mu_w C_w \gamma_e^2},$$

where

$G_P$ = Cumulative gas produced, SCF;
$B_{gi}$ = Initial gas formation volume factor, res. ft$^3$/SCF;
$B_g$ = Gas formation volume factor, res. ft$^3$/SCF;
$W_p$ = Cumulative water production, bbl;
$G$ = Original gas in place, SCF;
$B$ = Water influx constant or 6.28 $C_w \phi \gamma_w^2 h$, res. ft$^3$/psi;
$\Delta P_{avg}$ = Average reservoir pressure decrement, psi;
$Q_{T_D}$ = Dimensionless water influx function;
$T_D$ = Dimensionless time function;
$\phi$ = Average reservoir porosity, fraction;
$\mu$ = Water viscosity at reservoir condition;
$C_w$ = Water compressibility, vol/vol/psi;
$\gamma_e$ = Radius of the aquifer, ft;
$k$ = Average reservoir permeability, md;
$\Delta P$ = Pressure decrement, psi;
$t$ = Indicates time interval.

---

## PROGRAM DESCRIPTION

The program consists of a FORTRAN main program and two subprograms. Subprogram COMFAC is used to compute the gas compressibility factor. Subprogram SPLINE is a spline curve-fitting program, which is used to generate continuous values of the $Q_T$ function from user-supplied discrete values of dimensionless time, $T_D$, and the corresponding $Q_T$ function. Computed values of $X$ and $Y$ variables are then used to generate a least-squares straight line. The intercept of the straight line with the ordinate is then computed, and the value is printed as the original gas in place.

---

**Input Format**

| Card | Format | Content |
|------|--------|---------|
| 1 | 20A4 | A(I) |
| 2 | F5.3, F5.0, F5.2, F10.8, F10.1 | POR, PERMW, VISW, CW, RD |
| 3 | 3F5.0 | RT, PC, TC |
| 4 | I2 | N |
| 5–N | F5.1, F5.0, F15.5, F10.0, F5.3, F7.6 | C1(I), C2(I), C3(I), C4(I), C5(I), C6(I) |
| 6 | I2 | NN |
| 7–NN | 2F10.5 | TD(I), QT(I) |

---

**Description of Input Parameters**

$A(I)$ = Uses up to 80 alphanumeric characters for project identification;
POR = Average reservoir porosity, fraction;
PERMW = Average permeability in the aquifer, md;
VISW = Average reservoir water viscosity, cp;
CW = Average compressibility of reservoir water, vol/vol/psi;
RD = Radius of gas reservoir, ft;
RT = Reservoir temperature, °F;
PC = Critical pressure for the gas, psi;
TC = Critical temperature for the gas, °R;
N = Number of pressure-performance step history provided;
$C1(I)$ = Time, year;
C2 = Average reservoir pressure, psi;
C3 = Cumulative gas production, MMCF;
C4 = Cumulative water production, bbl;
C5 = Gas compressibility factor, Z;
C6 = Gas formation volume factor, res. ft$^3$/SCF;
$T_D$ = Dimensionless time;
$Q_T$ = Dimensionless water influx function;
NN = Number of discrete values of $T_D$ and $Q_T$ provided by the user.

**EXAMPLE PROBLEM**

Find the original gas in place for a water drive reservoir with the pressure-pro-duction-fluid property data shown in table 17.1. The following additional data are also available:

Reservoir temperature = 229° F,
Reservoir porosity = 0.209,
Reservoir permeability = 275 md,
Viscosity of water = 0.25 cp,
Water compressibility = $6.8 \times 10^6$ vol/vol/psi,
Gas drainage radius = 15983 ft,
Critical pressure = 1078 psi,
Critical temperature = 300° R,
Aquifer is infinite.

Table 17.2 presents the dimensionless time and the corresponding response function values to be used for the problem.

Table 17.1 Pressure, Production, and Fluid Property Data

| Time (year) | Average Reservoir Pressure (psia) | Cumulative Gas Production (MMCF) | Cumulative Water Production (bbl) | Z-Factor | Gas Formation Volume Factor (res. ft³/SCF) |
|---|---|---|---|---|---|
| 0 | 5392 | 0 | 0 | 1.053 | 0.003804 |
| 0.5 | 5368 | 677.746 | 3 | 1.052 | 0.003816 |
| 1 | 5292 | 2952.420 | 762 | 1.047 | 0.003854 |
| 1.5 | 5245 | 5199.568 | 2054 | 1.044 | 0.003878 |
| 2 | 5182 | 7132.759 | 3300 | 1.040 | 0.003911 |
| 2.5 | 5147 | 9196.910 | 4644 | 1.038 | 0.003930 |
| 3 | 5110 | 11171.520 | 5945 | 1.036 | 0.003949 |
| 3.5 | 5066 | 12999.530 | 7148 | 1.033 | 0.003971 |
| 4 | 5066 | 14769.528 | 8238 | 1.029 | 0.004002 |
| 4.5 | 4944 | 16316.950 | 9289 | 1.028 | 0.004008 |
| 5 | 4997 | 17867.981 | 10350 | 1.028 | 0.004007 |
| 5.5 | 4990 | 19416.030 | 11424 | 1.027 | 0.004010 |
| 6 | 4985 | 21524.783 | 12911 | 1.027 | 0.004018 |

**17.2 Dimensionless Time versus Dimensionless Response Function for an Infinite Aquifer**

| $T_D$ | $Q_T$ |
|-------|-------|
| 0 | 0 |
| 3.5 | 3.548 |
| 7 | 5.743 |
| 10.5 | 7.676 |
| 14 | 9.461 |
| 17.5 | 11.150 |
| 21 | 12.778 |
| 24.5 | 14.352 |
| 28 | 15.883 |
| 31.5 | 17.379 |
| 35 | 18.845 |
| 38.5 | 20.284 |
| 42 | 21.701 |

**INPUT LISTING**

FORTRAN CODING F

WW/CC   NAME
        DATE

PROGRAM
ROUTINE

O = ZERO      1 = ONE
Ø = ALPHA O   1 = ALPHA 1

| CARD | STATEMENT NUMBER (cols 1–50) |
|------|------------------------------|
| 1. | MCLEWEN RESERVOIR GULF COAST TEXAS |
| 2. | 27.51  .000001680  159.83.9 |
| 3. | 22.9  107.81  30.0 |
| 4. | 13 |
| 5-1 | 0.05  39.2  .6  0.1.053  .0038.09 |
| 5-2 | 0.55  36.8  6.7.746  3.1.052  .0038.16 |
| 5-3 | 1.05  29.2  29.52.420  7.62.1.047  .0038.54 |
| 5-4 | 1.55  24.5  51.99.56.8  2.05.4.1.044  .0038.78 |
| 5-5 | 2.05  18.2  71.312.75.9  3.00.1.040  .0039.11 |
| 5-6 | 2.55  47  91.96.91.0  46.44.1.038  .0039.30 |
| 5-7 | 3.05  10  111.71.52.9  59.45.1.036  .0039.49 |
| 5-8 | 3.55  06.6  12.99.9.53.0  71.48.1.033  .0039.71 |
| 5-9 | 4.05  006  147.69.52.8  82.38.1.029  .0040.02 |
| 5-10 | 4.55  99.4  163.16.95.0  92.89.1.028  .0040.08 |
| 5-11 | 5.05  99.7  178.67.98.1  103.50.1.028  .0040.07 |
| 5-12 | 5.55  990  194.16.030  114.24.1.027  .0040.10 |
| 5-13 | 6.05  98.5  215.24.78.3  129.11.1.027  .0040.13 |
| 6. | 13 |

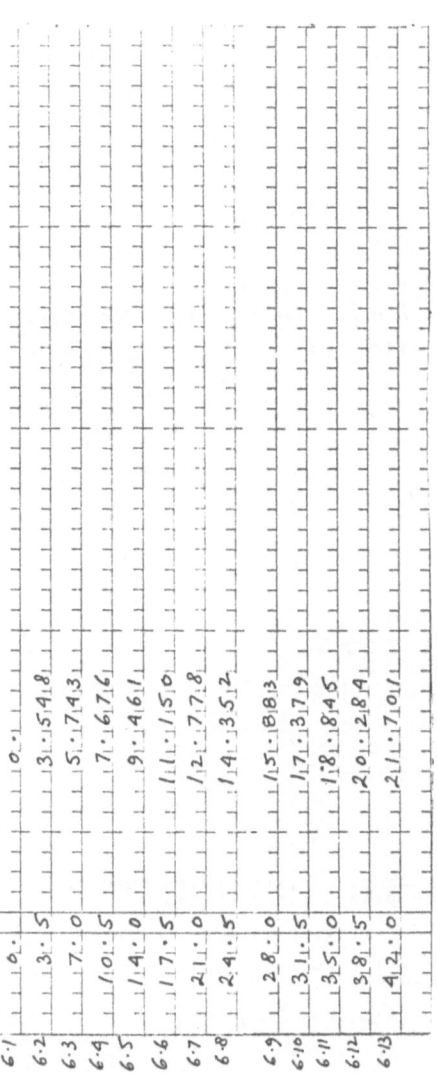

**SOURCE LISTING**

```
C     THIS PROGRAM CALCULATES ORIGINAL GAS-IN-PLACE FOR A WATER DRIVE
C     GAS RESERVOIR (RADIAL GEOMETRY) BY MATERIAL BALANCE METHOD
C
      DIMENSION C1(20),C2(20),C3(20),C4(20),C5(20),C6(20),C7(20),C8(20),
     *C9(20),C10(20),C11(20),C12(20),C13(20),C14(20),C15(20),C16(20)
      DIMENSION TD(50),QT(50),A(20)
      READ(5,1)(A(I),I=1,20)
    1 FORMAT(20A4)
      WRITE(6,2)
    2 FORMAT(//,3X,'DETERMINATION OF ORIGINAL GAS-IN-PLACE BY MBE FOR'
     *,' WATER DRIVE GAS-RESERVOIR')
      WRITE(6,3)(A(I),I=1,20)
    3 FORMAT(/,3X,20A4)
      READ(5,4)POR,PERMW,VISW,CW,RD
    4 FORMAT(F5.3,F5.0,F5.2,F10.8,F10.1)
      READ(5,5)RT,PC,TC
    5 FORMAT(3F5.0)
C
C     POR=HYDROCARBON POROSITY
C     C1=TIME IN YEARS
C     C2=RESERVOIR PRESSURE, PSIA
C     C3=CUMULATIVE GAS PRODUCTION, MMSCF
C     C4=CUMULATIVE WATER PRODUCTION, BBL
C     C5=GAS-DEVIATON FACTOR
C     C6=GAS FORMATION VOLUME FACTOR, RES.CF/SCF
C
      TDF=2.31045*PERMW/(POR*VISW*CW*RD**2)
      READ(5,115)N
      DO 100 I=1,N
      READ(5,101)C1(I),C2(I),C3(I),C4(I),C5(I),C6(I)
  101 FORMAT(F5.1,F5.0,F15.5,F10.0,F5.3,F7.6)
      IF(C5(I).EQ.0.0.AND.C6(I).EQ.0.0)CALL COMFAC(C1(I),RT,PC,TC,Z)
      IF(C5(I).EQ.0.0)C5(I)=Z
      IF(C6(I).EQ.0.0)C6(I)=19.48*C5(I)/C2(I)
  100 CONTINUE
      DO 110 I=2,N
      C7(I)=C6(I)-C6(1)
      C8(I)=C3(I)*C6(I)
      C9(I)=C8(I)+5.615*C4(I)*1.0E-6
      C10(I)=C9(I)/C7(I)
  110 CONTINUE
      READ(5,115)NN
  115 FORMAT(I2)
      DO 120 I=1,NN
      READ(5,121)TD(I),QT(I)
  121 FORMAT(2F10.5)
  120 CONTINUE
      C13(1)=0.
      DO 130 I=2,N
      C11(I)=TDF*C1(I)
      CALL SPLINE(TD,QT,NN)
      CALL GRATA(C11(I),AA,NN)
      CALL TERPA(C11(I),TX,NN)
      C12(I)=TX
      C13(I)=C2(I-1)-C2(I)
      C14(I)=0.5*(C13(I-1)+C13(I))
  130 CONTINUE
      DO 140 J=2,N
      SUM=0.
      DO 141 I=2,J
      SUM=SUM+C14(I)*C12(J+2-I)
  141 CONTINUE
      C15(J)=SUM
  140 CONTINUE
      DO 142 I=2,N
      C16(I)=C15(I)/C7(I)
  142 CONTINUE
      WRITE(6,145)
  145 FORMAT(/,3X,'TIME',3X,'CUM.GAS PROD.',3X,'CUM WATER PROD.',
     *3X,'RESER. PRESS.',3X,'VALUE OF Y',3X,'VALUE OF X')
      WRITE(6,146)
  146 FORMAT(4X,'YR',9X,'MMCF',13X,'BBL',13X,'PSIA',11X,'MMCF')
      DO 150 I=2,N
      WRITE(6,151)C1(I),C3(I),C4(I),C2(I),C10(I),C16(I)
  151 FORMAT(3X,F4.0,F13.0,F16.0,F17.0,F16.0,F13.0)
  150 CONTINUE
      SUMX=0.
      SUMY=0.
      SUMXY=0.
      SUMX2=0.
```

```
      SUMY2=0.
      XN=N-1
      DO 160 I=2,N
      SUMX=SUMX+C16(I)
      SUMX2=SUMX2+C16(I)**2
      SUMY=SUMY+C10(I)
      SUMY2=SUMY2+C10(I)**2
      SUMXY=SUMXY+C10(I)*C16(I)
  160 CONTINUE
      CEPT=(SUMX2*SUMY-SUMX*SUMXY)/(XN*SUMX2-SUMX**2)
      SLOPE=(SUMY-XN*CEPT)/SUMX
      WRITE(6,161) CEPT
  161 FORMAT(//,3X,'ORIGINAL OIL IN PLACE',3X,F15.0,1X,'MMCF')
      WRITE(6,162)SLOPE
  162 FORMAT(//,3X,'SLOPE OF THE LINE',24X,F15.3,3X,'MM RESERVOIR',
     *' C.FT/PSI')
      STOP
      DEBUG SUBCHK
      END
      SUBROUTINE COMFAC(T,S,PC,TC,Z)
      PR=T
      X=S+460.
      RRT=TC/X
      RP=PR/PC
      A=0.06125*RRT*EXP(-1.2*(1.-RRT)**2)
      B=RRT*(14.76-9.76*RRT+4.58*RRT**2)
      C=RRT*(90.7-242.2*RRT+42.4*RRT**2)
      D=2.18+2.82*RRT
      Y=0.01
      DO 2 J=1,30
      IF(Y.GT.1.)Y=0.6
      F=-A*RP+Y*(1.+(Y*(1.+Y*(1.-Y))))/(1.-Y)**3-3*Y**2+C*Y**D
      IF(ABS(F)-1.0E-6)4,4,3
    3 DFDY=(1.+4.*Y*(1.+Y*(1.-Y))+Y**4)/(1.-Y)**4-2.*B*Y+D*C*Y**(D-1.)
    2 Y=Y-F/DFDY
    4 Z=A*RP/Y
      RETURN
      END
      SUBROUTINE SPLINE(XI,YI,N)
      DIMENSION W(99),Q(99),B(99),A(99),C(99),S(99),Z(99)
      DIMENSION XI(50),YI(50),X(99),Y(99)
      DATA A(1),C(1),Z(1)/-1.0.0.,0./
      DO 11 I=1,N
      X(I)=XI(I)
      Y(I)=YI(I)
   11 CONTINUE
      DO 1 J=2,N
      W(J)=X(J)-X(J-1)
    1 CONTINUE
      NM=N-1
      DO 2 J=2,NM
      WJ=W(J)
  160 WP=W(J+1)
      WS=WJ+WP
      QJ=WJ/WS
      Q(J)=QJ
      QA=1.-.5*QJ*A(J-1)
      A(J)=.5*(1.-QJ)/QA
      B(J)=3.*(WP*Y(J-1)-WS*Y(J)+WJ*Y(J+1))/WP/WJ/WS
      C(J)=(B(J)-.5*QJ*C(J-1))/QA
    2 CONTINUE
      S(N)=C(NM)/(1.+A(NM))
      S(NM)=S(N)
      NMM=N-2
      DO 3 JJ=1,NMM
      J=NMM-JJ+1
      S(J)=C(J)-A(J)*S(J+1)
    3 CONTINUE
      DO 4 J=2,N
      Z(J)=Z(J-1)+.5*W(J)*(Y(J)+Y(J-1)-.0825*W(J)**2*(S(J)+S(J-1)))
    4 CONTINUE
      RETURN
C
      ENTRY DERIVA(XV,YV,N)
      DO 5 JJ=2,N
      J=JJ
      IF(XV.GT.X(J))GOTO 5
      GOTO 6
    5 CONTINUE
    6 WJ=W(J)
      D1=(XV-X(J-1))/WJ
      D2=(X(J)-XV)/WJ
      YV=(Y(J)-Y(J-1))/WJ+.5*WJ*((D1*D1-.333333)*S(J)
     *-(D2*D2-.333333)*S(J-1))
      RETURN
      ENTRY TERPA(XV,YV,N)
      DO 7 JJ=2,N
      J=JJ
```

```
      IF(XV.GT.X(J)) GOTO 7
      GOTO 8
    7 CONTINUE
    8 WJ=W(J)
      D1=(XV-X(J-1))/WJ
      D2=(X(J)-XV)/WJ
      D3=WJ*WJ/6.
      YV=D1*(Y(J)+D3*(D1*D1-1.)*S(J))
      YV=YV+D2*(Y(J-1)+D3*(D2*D2-1.)*S(J-1))
      RETURN
      ENTRY GRATA(XV,YV,N)
      DO 9 JJ=2,N
      J=JJ
      IF(XV.GT.X(J)) GOTO 9
      GOTO 10
    9 CONTINUE
   10 WJ=W(J)
      D1=((XV-X(J-1))/WJ)**2
      D2=((X(J)-XV)/WJ)**2
      D3=.0825*WJ*WJ
      YV=Z(J-1)+.5*WJ*(D1*(Y(J)+D3*(D1-2.)*S(J))+(1.-D2)*(Y(J-1)+D3*(D2
     *-1.)*S(J-1)))
      RETURN
      END
//GO.SYSIN DD *
MCEWEN RESERVOIR          GULF COAST      TEXAS            P 217
 .209 275.  .25 .00000680    15983.35
 229.1078. 300.
13
   0.05392.       0.               0.    1.053.003804
   0.55368.       677.746          3.    1.052.003816
   1.05292.       2952.420         762.  1.047.003854
   1.55245.       5199.568         2054. 1.044.003878
   2.05182.       7132.759         3300. 1.040.003911
   2.55147.       9196.910         4644. 1.038.003930
   3.05110.       11171.529        5945. 1.036.003949
   3.55066.       12999.530        7148. 1.033.003971
   4.05006.       14769.528        8238. 1.029.004002
   4.54994.       16316.950        9289. 1.028.004008
   5.04997.       17867.981        10350.1.028.004007
   5.54990.       19416.030        11424.1.027.004010
   6.04985.       21524.783        12911.1.027.004013
13
   0.             0.
   3.5            3.548
   7.             5.743
   10.5           7.676
   14.0           9.461
   17.5           11.150
   21.0           12.778
   24.5           14.352
   28.            15.883
   31.5           17.379

   35.0           18.845
   38.5           20.284
   42.0           21.701
 /*
```

## OUTPUT LISTING

```
DETERMINATION OF ORIGINAL GAS-IN-PLACE BY MBE FOR WATER DRIVE GAS-RESERVOIR

MCEWEN RESERVOIR          GULF COAST      TEXAS           P 217
```

| TIME<br>YR | CUM.GAS PROD.<br>MMCF | CUM WATER PROD.<br>BBL | F ESER. PRESS.<br>PSIA | VALUE OF Y<br>MMCF | VALUE OF X |
|---|---|---|---|---|---|
| 1. | 678. | 3. | 5368. | 215527. | 3548057. |
| 1. | 2952. | 762. | 5292. | 227658. | 4926350. |
| 2. | 5200. | 2054. | 5245. | 272641. | 8073883. |
| 2. | 7133. | 3300. | 5182. | 260886. | 9772659. |
| 3. | 9197. | 4644. | 5147. | 287066. | 12449722. |
| 3. | 11172. | 5945. | 5110. | 304478. | 14648223. |
| 4. | 13000. | 7148. | 5066. | 339350. | 16429913. |
| 4. | 14770. | 8238. | 5006. | 298755. | 17496480. |
| 5. | 16317. | 9289. | 4994. | 320838. | 20648848. |
| 5. | 17868. | 10350. | 4997. | 352980. | 24107472. |
| 6. | 19416. | 11424. | 4990. | 378265. | 26913296. |
| 6. | 21525. | 12911. | 4985. | 413647. | 29615536. |

```
ORIGINAL OIL IN PLACE            197747. MMCF

SLOPE OF THE LINE                              0.007   MM RESERVOIR C.FT/PSI
```

### BIBLIOGRAPHY

Craft, B. C., and Hawkins, M. F.: *Applied Petroleum Reservoir Engineering,* Prentice-Hall, Englewood Cliffs, N.J. (1959).

Havlena, D., and Odeh, A. S.: "The Material-Balance as an Equation of a Straight Line," *J. Pet. Tech.* (August 1963) 896–900.

———: "The Material-Balance as an Equation of a Straight Line—Part II : Field Cases," *J. Pet. Tech.* (July 1964) 815–822.

# DETERMINATION OF ORIGINAL GAS IN PLACE FOR AN ABNORMALLY PRESSURED RESERVOIR

## PURPOSE

The program uses the material balance method to determine the original gas in place of an abnormally pressured, essentially volumetric gas reservoir from the available pressure and production data.

The material balance equation used is designed to take into account information such as rock, shale, and water compressibilities as well as size of any aquifer (small) associated with the reservoir. A conventional gas material balance equation is developed by assuming that rock, shale, and water compressibilities are negligible compared to the compressibility of the gas. This assumption is realistic for a normally pressured volumetric gas reservoir, especially where the reservoir rock is consolidated. However, when the reservoir pressures are high, the gas compressibility ($\approx 1/P$) values are low and comparable in order of magnitude with the rock and water compressibilities and are not negligible. Hence, the application of a conventional material balance approach ($P/Z$ versus $G_P$) to an abnormally pressured reservoir with unconsolidated rock characteristics tends to provide a higher value of estimated original gas in place from the early production data.

## METHOD

$$C_e = C_r(1 + R) + C_w(S_w + R) + C_s(1 + R)\frac{f}{\phi} \tag{18.1}$$

$$G_p = \frac{T_{sc}P_iV_i}{TP_{sc}Z_i} - \frac{T_{sc}V_i}{TP_{sc}}\left(1 - \frac{C_eP_i}{1 - S_w}\right)\frac{P}{Z} - \frac{T_{sc}V_i}{TP_{sc}}\left(\frac{C_e}{1 - S_w}\right)\frac{P^2}{Z} \tag{18.2}$$

or $G_p = m\left\{\dfrac{P_i}{Z_i} - \left(1 - \dfrac{C_e P_i}{1 - S_w}\right)\dfrac{P}{Z} - \left(\dfrac{C_e}{1 - S_w}\right)\dfrac{P^2}{Z}\right\}$      (18.3)

or $G_P = mx$

$m = \dfrac{T_{sc} V_i}{TP_{sc}};$

$x = \dfrac{P_i}{Z_i} - \left(1 - \dfrac{C_e P_i}{1 - S_w}\right)\dfrac{P}{Z} - \left(\dfrac{C_e}{1 - S_w}\right)\dfrac{P^2}{Z};$      (18.4)

where

$C_r$ = Rock compressibility, vol/vol/psi;
$R$ = Size of the associated aquifer, multiple of the reservoir size;
$C_w$ = Water compressibility, vol/vol/psi;
$S_w$ = Interstitial water saturation, fraction;
$C_s$ = Bulk compressibility of shale, pore volume/bulk volume/psi;
$f$ = Shale fraction of the reservoir and aquifer volume;
$\phi$ = Porosity, fraction;
$C_e$ = Effective compressibility, vol/vol/psi;
$G_P$ = Cumulative gas produced, MMSCF;
$T_{sc}$ = Standard temperature, °R;
$P_i$ = Initial reservoir pressure, psi;
$V_i$ = Initial hydrocarbon pore volume, MMCF;
$T$ = Reservoir temperature, °R;
$P_{sc}$ = Standard pressure, psi;
$Z_i$ = Initial gas compressibility factor;
$P$ = Reservoir pressure, psi;
$Z$ = Gas compressibility factor at pressure $P$ and reservoir temperature.

Equation (18.4) is an equation of a straight line; hence, if a coordinate plot of $G_P$ versus $X$ is made, one should obtain a straight line passing through the origin. Extrapolation of this line to an abscissa value of $P_i/Z_i$ will provide a corresponding ordinate value that is the estimated value of the original gas in place.

---

**PROGRAM DESCRIPTION**

The program consists of a FORTRAN main program and a subprogram COM-FAC. COMFAC is used to determine the gas compressibility factor at various reservoir pressures and temperature conditions.

The user supplies a set of pressure-cumulative production ($G_P$) data. At each of these data points, corresponding $X$ values are computed, and the slope of a least-squares straight line through the points is evaluated. Using this slope and assuming the reservoir pressure to be zero, equation (18.3) is solved to obtain the original gas-in-place value. The intercept value is also computed and printed. If the computed intercept value is negligible compared to the estimated gas-in-place value, then the computation as well as the input data validity are justified. However, if the intercept is significant compared to the original gas in place, then the assumption regarding input data must be verified and checked. Often the aquifer size is not accurately known. Therefore, if other reservoir and rock properties are accurate, then one may wish to make various assumptions regarding aquifer size and make a series of runs in order to arrive at a reasonable estimate of the intercept.

## Input Format

| Card | Format | Content |
|------|--------|---------|
| 1 | 20A4 | A(J) |
| 2 | 20A4 | B(I) |
| 3 | I2 | N |
| 4 | F6.0, F4.0, F6.0 | PRI, TRE, ABP |
| 5 | F3.2, 3F8.7, F4.2, 3F4.3 | R, RC, WC, SC, SF, POR, SWI, GG |
| 6–N | F10.0 | GP(I), P(I) |

## Description of Input Parameters

A(J) = Uses up to 80 alphanumeric characters for project identification;
B(I) = Uses up to 80 alphanumeric characters for project identification;
N = Number of pressure-production data points provided by the user;
PRI = Initial reservoir pressure, psi;
TRE = Reservoir temperature, °F;
ABP = Reservoir abandonment pressure, psi;
R = Aquifer size, multiple of reservoir size;
RC = Rock compressibility, vol/vol/psi;
WC = Water compressibility, vol/vol/psi;
SC = Bulk compressibility of shale, bulk vol/pore vol/psi;
SF = Shale fraction in the reservoir and aquifer;
POR = Reservoir porosity, fraction;
SWI = Interstitial water saturation, fraction;
GG = Gas gravity (air = 1);
GP(I) = Cumulative gas produced at reservoir pressure P(I), MMSCF;
P(I) = Reservoir pressure, psi.

**EXAMPLE PROBLEM**

The early pressure-production data for an abnormally pressured reservoir are given in table 18.1. Other reservoir and fluid data are given below. Find the original gas in place and recoverable reserve if the abandonment pressure is 2000 psi.

Reservoir depth = 8000 ft,
Reservoir temperature = 230° F,
Size of the aquifer = 0.5 times the reservoir volume,
Rock compressibility = 4 × $10^{-6}$ vol/vol/psi,
Water compressibility = 3 × $10^{-6}$ vol/vol/psi,
Bulk shale compressibility = 11.3 × $10^{-6}$ bulk vol/pore vol/psi,
Shale fraction = 22%,
Average reservoir porosity = 24%,
Interstitial water saturation = 34%,
Gas gravity = 0.7 (air = 1).

**Table 18.1 Pressure and Cumulative Production**
**History of an Abnormally Pressured**
**Reservoir**

| Pressure (psi) | Cumulative Production (MMSCF) |
|---|---|
| 8921 | 0 |
| 8845 | 552 |
| 8322 | 4409 |
| 7417 | 11275 |
| 6838 | 15862 |
| 6064 | 22288 |
| 5490 | 27334 |
| 4781 | 33983 |
| 4104 | 40883 |

**INPUT LISTING**

# FORTRAN CODING FORM

UU/CC   NAME
        DATE

PROGRAM
ROUTINE

0 = ZERO   1 = ONE    2 = TWO
Ø = ALPHA O   I = ALPHA I   Z = ALPHA Z

| CARD | STATEMENT NUMBER 1,2,3,4,5 | 6 | 7,8,9,10,11,12,13... |
|---|---|---|---|
| 1. | PRØBLEM #16 | | MURRAY VITAM PL-121/I |
| 2. | WELL #6 | | FIELD SIINHA RESERVØIR RAHUL |
| 3. | 9 | | |
| 4. | 8921 . 2100 | | |
| 5. | 501000004010000000130 0000III3 122 2401 3401 7000 | | |
| 6-1 | 10 8921 | | |
| 6-2 | 552 8845 | | |
| 6-3 | 4 409 8322 | | |
| 6-4 | 11 275 7417 | | |
| 6-5 | 15 862 6838 | | |
| 6-6 | 22 288 6064 | | |
| 6-7 | 27 334 5490 | | |
| 6-8 | 33 983 4781 | | |
| 6-9 | 4 08 883 4104 | | |

## SOURCE LISTING

```
C          THIS PROGRAM CALCULATES INITIAL GAS IN PLACE AND GAS RECOVERY BY
C          MATERIAL BALANCE APPROACH FOR VOLUMETRIC ABNORMALLY PRESSURED
C          RESERVOIR WHERE ROCK AND WATER COMPRESSIBILITIES ARE NOT
C          NEGLIBLE.  ALSO SHALE COMPRESSIBILITIES ARE TAKEN INTO
C          CONCIDERATION.  ASSOCIATED ACQUIFER SIZE, IF PRESENT SHOULD
C          BE CONSIDERED.
C
           DIMENSION A(20),B(20),GP(20),P(20),X(20),Y(20),POZ(20)
           READ(5,1)(A(I),I=1,20)
           READ(5,1)(B(I),I=1,20)
    1 FORMAT(20A4)
           WRITE(6,2)(A(I),I=1,20)
           WRITE(6,2)(B(I),I=1,20)
    2 FORMAT(3X,20A4)
           WRITE(6,3)
    3 FORMAT(//,1X,'DETERMINATION OF GAS IN PLACE AND RECOVERY FACTOR',
       */,3X,'VOLUMETRIC RESERVOIR WITH ROCK AND WATER COMPRESSIBILITIES',
       */,3X,'CONSIDERED')
C          N=NUMBER OF PRESSURE PRODUCTION DATA AVAILABLE
           READ(5,7)N
           READ(5,6)PRI,TRE,ABP
           READ(5,5)R,RC,WC,SC,SF,POR,SWI,GG
C          RC=ROCK COMPRESSIBILITY, VOL/VOL/PSI
C          WC=WATER COMPRESSIBILITY, VOL/VOL/PSI
C          SC=SHALE COMPRESSIBILITY, VOL/VOL/PSI
C          R=SIZE OF ASSOCIATED AQUIFER, FRACTION OF THE RESERVOIR VOLUME
C          SF=SCHALE FRACTION OF THE RESERVOIR ROCK, FRACTION
C          POR=EFFECTIVE POROSITY OF THE RESERVOIR ROCK FRACTION
C          SWI=WATER SATURATION, FRACTICN
C          GG=GAS GRAVITY (AIR=1.0)
           DO 10 I=1,N
           READ(5,11)GP(I),P(I)
C          GP=CUMULATIVE PRODUCTION, MMCF
C          P=RESERVOIR PRESSURE, PSI
   10 CONTINUE
           TR=TRE+460.
           CE=RC*(1.+R)+WC*(SWI+R)+SC*(1.+R)*SF/POR
           PC=676.2366+6.9896*GG-25.8273*GG**2
           TC=142.7712+380.9671*GG-36.4051*GG**2
           CALL COMFAC(PRI,TR,PC,TC,Z)



           ZI=Z
           DO 20 I=1,N
           CALL CCMFAC(P(I),TR,PC,TC,Z)
           X(I)=PRI/ZI-(P(I)/Z)*(1.-CE*PRI/(1.-SWI))-(P(I)**2/Z)
       **(CE/(1.-SWI))
           Y(I)=GP(I)
           POZ(I)=P(I)/Z
   20 CONTINUE
           SUMX=0.
           SUMY=0.
           SUMX2=0.
           SUMY2=0.
           SUMXY=0.
           DO 50 I=1,N
           SUMX=SUMX+X(I)
           SUMY=SUMY+Y(I)
           SUMX2=SUMX2+X(I)**2
           SUMY2=SUMY2+Y(I)**2
           SUMXY=SUMXY+X(I)*Y(I)
   50 CONTINUE
           SLOPE=(SUMXY-SUMX*SUMY/FLOAT(N))/(SUMX2-SUMX**2/FLOAT(N))
           CEPT=(SUMY*SUMX2-SUMX*SUMXY)/(FLOAT(N)*SUMX2-SUMX**2)
           OGIP=SLOPE*PRI/ZI
           CALL COMFAC(ABP,TR,PC,TC,Z)
           XX=PRI/ZI-(ABP/Z)*(1.-(CE*PRI)/(1.-SWI))-(ABP**2/Z)*CE/(1.-SWI)
           RR=SLOPE*XX
           ABPZ=ABP/Z
           RF=(RR/OGIP)*100.
   11 FORMAT(F10.0,F6.0)
           WRITE(6,13)
   13 FORMAT(//,3X,'RESERVOIR',8X,'CUMMULATIVE')
           WRITE(6,14)
   14 FORMAT(4X,'PRESSURE',9X,'PRODUCTION')
           WRITE(6,15)
   15 FORMAT(6X,'PSIA',14X,'MMCF')
           DO 70 I=1,N
   70 WRITE(6,71)P(I),GP(I),X(I),Y(I),POZ(I)
```

```
   71  FORMAT(2X,F8.0,10X,3(F10.0,3X),3X,F10.0)
       WRITE(6,60)OGIF
   60  FORMAT(//,3X,'ORIGINAL GAS IN PLACE : ',F15.0,3X,'MMCF')
       WRITE(6,61)RR,AEP
   61  FORMAT(/,3X,'RECCVERABLE RESERVE : ',F15.0,3X,'MMCF',1X,'A',
      *F6.0,1X,'PSIA',1X,'ABANDONMENT')
       WRITE(6,51)CEPT
   51  FORMAT(3X,'CEPT = ',F10.0)
       WRITE(6,62)RF
   62  FORMAT(/,3X,'RECCVERY',F7.2,1X,'PERCENT'/)
    6  FORMAT(F6.0,F4.0,F6.0)
    5  FORMAT(F3.2,3F8.7,F4.2,3F4.3)
    7  FORMAT(I2)
       STOP
       END
       SUBROUTINE COMFAC(T,S,PC,TC,Z)
       PR=T
       X=S
       RRT=TC/X
       RP=PR/PC
       A=0.06125*RRT*EXP(-1.2*(1.-RRT)**2)
       B=RRT*(14.76-9.76*RRT+4.58*RRT**2)
       C=RRT*(90.7-242.2*RRT+42.4*RRT**2)
       D=2.18+2.82*RRT
       Y=0.01
       DO 2 J=1,30
       IF(Y.GT.1.)Y=0.6
       F=-A*RP+Y*(1.+(Y*(1.+Y*(1.-Y))))/(1.-Y)**3-B*Y**2+C*Y**D
       IF(ABS(F)-1.0E-6)4,4,3
    3  DFDY=(1.+4.*Y*(1.+Y*(1.-Y))+Y**4)/(1.-Y)**4-2.*B*Y+D*C*Y**(D-1.)
    2  Y=Y-F/DFDY
    4  Z=A*RP/Y
       RETURN
       END
```

```
PROBLEM # 16          MURRAY          UTAH                        P 211
WELL # 16             FIELD    SINHA         RESERVOIR RAHUL
  9
 8921.230. 2000.
.50.0000040.0000030.0000113 .22.240.340.700
       0. 8921.
     552. 8845.
    4409. 8322.
   11275. 7417.
   15862. 6838.
   22288. 6064.
   27334. 5490.
   33983. 4781.
   40883. 4104.
/*
/*.
```

## OUTPUT LISTING

```
PROBLEM # 16        MURRAY        UTAH                            P 211
WELL # 16           FIELD    SINHA           RESERVOIR RAHUL

DETERMINATION OF GAS IN PLACE AND RECOVERY FACTOR
    VOLUMETRIC RESERVOIR WITH ROCK AND WATER COMPRESSIBILITIES
    CONSIDERED

RESERVOIR           CUMMULATIVE
PRESSURE            PRODUCTION
  PSIA                MMCF
 8921.                  0.              0.              0.        6715.
 8845.                552.             42.            552.        6691.
 8322.               4409.            335.           4409.        6523.
 7417.              11275.            858.          11275.        6197.
 6838.              15862.           1207.          15862.        5961.
 6064.              22288.           1696.          22288.        5602.
 5490.              27334.           2080.          27334.        5298.
 4781.              33983.           2586.          33983.        4862.
 4104.              40883.           3111.          40883.        4371.

ORIGINAL GAS IN PLACE :              88228.   MMCF

RECOVERABLE RESERVE :                66026.   MMCF A 2000. PSIA ABANDONMENT
CEPT =         4.

RECOVERY   74.84 PERCENT
```

## BIBLIOGRAPHY

Bourgoyne, A. T., Hawkins, M. F., Lavagnial, F. P., and Wickenhauser, T. L.: "Shale
    Water as a Pressure Support Mechanism in Superpressure Reservoir," paper SPE 3851
    presented at the Abnormal Subsurface Pressure Meeting, May 15–16, 1972, Baton
    Rouge, Louisiana.

# 19 STATIC/FLOWING BOTTOMHOLE PRESSURE FOR A GAS WELL

**PURPOSE**

The program converts static/flowing wellhead pressure to bottomhole (sand face) pressure using a stepwise numerical procedure based on the Interstate Oil Compact Commission (IOCC) manual. Either the tubular or annular flow option may be selected for the flowing pressure calculations.

**METHOD**

$$P_{BH} = P_{WH} + \Delta P,$$

$$\Delta P = \frac{3(37.5GH)}{I_1 + 4I_2 + I_3}.$$

Subscripts 1, 2, and 3 indicate conditions at the wellhead, at $H/2$ and $H$ respectively.

$$I_n = \frac{P/TZ}{(P/TZ)^2/1000 + (F_r Q_m)^2}$$

$$F_r = \frac{0.10797}{d_e^{2.612}} \qquad \text{for } d_e > 4.277 \text{ in}$$

$$F_r = \frac{0.10337}{d_e^{2.582}} \qquad \text{for } d_e \leq 4.277 \text{ in}$$

$$d_e^{2.612} = (d_i - d_o)^{1.612}(d_i + d_o)$$

$$d_e^{2.582} = (d_i - d_o)^{1.582}(d_i + d_o)$$

183

where

$P_{BH}$ = Bottomhole pressure, psia;
$P_{WH}$ = Wellhead pressure, psig;
$G$ = Gas gravity (air = 1);
$H$ = Length of the flow string, ft;
$F_r$ = Friction factor;
$T$ = Absolute temperature, °R;
$Z$ = Gas compressibility factor;
$Q_m$ = Flow rate, MMCFD (14.65 psia and 60° F);
$d_e$ = Equivalent diameter of the flow channel, in;
$d_i$ = Inside pipe diameter, in (tubing for tubular flow, casing for annular flow);
$d_o$ = Outside pipe diameter, in (tubing for annular flow);
$I_n$ = Iterative value at the $n^{th}$ iteration.

For annular flow, $d_o$ = External diameter of the tube;
For annular flow, $d_i$ = Internal diameter of casing;
For tubular flow, $d_i$ = Internal diameter of tubing;
For tubular flow, $d_o$ must be zero.

## PROGRAM DESCRIPTION

The program consists of a FORTRAN main program and two subprograms. Sub-program COMFAC generates the gas compressibility factor values, while sub-program BHPRS computes the bottomhole pressure. The static or flowing column option controls the type of calculation desired; that is, if the selected option is for static column calculation and a flow rate value has been provided, then the program automatically sets the flow rate to zero. Conversely, if the selected option is for flowing column and the flow rate as provided by the user is zero, static column pressure will be computed with a warning message ("flow rate is zero; therefore, desired flowing bottomhole pressure (BHP) cannot be calculated").

Similarly, if annular flow is desired, the internal casing diameter must be a nonzero positive number. If casing internal diameter is given as zero, the program will assume tubular flow.

### Input Format

| Card | Format | Content |
|------|--------|---------|
| 1 | 20A4 | A(I) |
| 2 | I2, F5.1, F4.0, F6.0, 2F6.3 | N1, BO, TR, DE, DI, DI2 |
| 3 | 2F5.0, 4F5.3 | PC, TC, C1, C2, C3, GG |
| 4–N1 | F5.0, F5.1, F10.1, I1 | PWF(I), TF(I), Q(I), N2 |

### Description of Input Parameters

$A(I)$ = Uses up to 80 alphanumeric characters for project identification;

$N1$ = Number of calculations desired;

$BO$ = Barometric pressure, psia;

$TR$ = Reservoir temperature, °F;

$DE$ = Length of the flow string as well as the vertical depth, ft (no hole deviation is considered);

$DI$ = Internal diameter of the tubing for tubular flow, in;

   = External diameter of the tubing for annular flow, in;

$DI2$ = Internal diameter of the casing for annular flow, in; $DI2 = 0$ for tubular flow;

$PC$ = Critical pressure of the gas, psi;

$TC$ = Critical temperature of the gas, °R;

$C1$ = Molecular fraction of carbon dioxide in the gas;

$C2$ = Molecular fraction of nitrogen in the gas;

$C3$ = Molecular fraction of hydrogen sulfide in the gas;

$GG$ = Gas gravity (air = 1);

$PWF(I)$ = Wellhead pressure, psig;

$TF(I)$ = Wellhead temperature, °F;

$Q(I)$ = Flow rate, MCFD;

$N2$ = Calculation option: $N2 = 1$, static column; $N2 = 2$, flowing column.

### EXAMPLE PROBLEM

Wellhead pressures for a gas well under flowing and static conditions along with flow rates are given in table 19.1. Other available information is given in the following. Determine corresponding bottomhole pressures.

Vertical depth of the flow string = 8226 ft

Reservoir temperature = 250° F

**19.1 Wellhead Pressure and Flow Rates**

| Wellhead Pressure (psig) | Flow Rate (MCFD) | Wellhead Temperature (°F) |
|---|---|---|
| 2272 | 4803 | 78 |
| 2260 | 5000 | 78 |
| 2000 | 5500 | 78 |
| 2954 | 0 | 60 |
| 2900 | 0 | 60 |
| 2850 | 0 | 60 |

Gas gravity = 0.761
Critical pressure = 662 psi
Critical temperature = 398° R

The well flows through the tubing string.

**INPUT LISTING**

# FORTRAN CODING FORM

UW/CC   NAME _____   DATE _____

PROGRAM _____
ROUTINE _____

0 = ZERO   1 = ONE   2 = TWO
Ø = ALPHA O   1 = ALPHA I   2 = ALPHA Z

| STATEMENT NUMBER 1,2,3,4,5 | 6 | 7,8,9,10,11...59 |
|---|---|---|
| CARD | | |
| 1. PRØBL | Ø | Em #1/2A    GREENFIELD NEILL #1  SALTLAKE CITY  UTIAH |
| 2. 61/12. | | 2,50. 8,26. 2.41 0.0 |
| 3. 6612. | | 3981. 01.01 01.01 01.7161 |
| 4-1 2,2,72. | | 78.0 4803. 2 |
| 4-2 2,26,0. | | 78.0 5000. 2 |
| 4-3 2,00,0. | | 78.0 5500. 2 |
| 4-4 2,9,5,4. | | 60.0 0.1 1 |
| 4-5 2,90,0. | | 60.0 0.1 |
| 4-6 2,85,0. | | 60.0 0.1 |

## SOURCE LISTING

```
C       THIS PROGRAM IS USED TO DETERMINE BOTTOM-HOLE STATIC AND FLCWING
C       PRESSURE FROM WELL-HEAD PRESSURE DATA FOR A GAS WELL.
C
        DIMENSION Q(20),PWF(20),PBHF(20),A(20),TF(20),N2(20)
        READ(5,3)(A(I),I=1,20)
      3 FORMAT(20A4)
        READ(5,1)N1,BO,TR,DE,DI,DI2

      1 FORMAT(I2,F5.1,F4.0,F6.0,2F6.3)
C       N1=NUMBER OF CALCULATIONS TO BE PERFCRMED
C       N2=PRESSURE CODE; N2=1 STATIC PRESSURE; N2=2 FLOWING PRESSURE
C       BO=BAROMETRIC PRESSURE
C       TR=RESERVOIR PRESSURE
C       TR=RESERVOIR TEMPERATURE, DEG. F
C       DE=DEPTH
C       DI=INTERNAL DIAMETER OF TUBE(IF FLOWING THRU TUBE), INCHES
C       DI=EXTERNAL DIAMETER OF TUBE(IF FLCWING THRU ANNULUS), INCHES
C       DI2=INTERNAL DIAMETER OF CASING(IF ANNULAR FLOW), INCHES,
C       OTHERWISE ZERO.
        READ(5,2)PC,TC,C1,C2,C3,GG
      2 FORMAT(2F5.0,4F5.3)
C       PC=CRITICAL PRESSURE OF THE GAS, PSIA
C       TC=CRITICAL TEMPERATURE OF GAS, DEG RANKINE
C       C1,C2,C3=MOL FRACTIONS
        DC 100 I=1,N1
        READ(5,5)PWF(I),TF(I),Q(I),N2(I)
    100 CONTINUE
C       PWF=WELL HEAD PRESSURE, PSIG
C       TF=WELL HEAD TEMPERATURE, DEG FARENHEIT
C       Q=FLOW RATE, MCF, ZERO FOR STATIC CONDITIONS
C       N2=PRESSURE CODE, 1 FOR STATIC, 2 FOR FLCWING
      5 FORMAT(F5.0,F5.1,F10.1,I1)
        TRA=TR+460.
        DO 200 I=1,N1
        PWF(I)=PWF(I)+BO
        TF(I)=TF(I)+460.
        IF(N2(I).EQ.1.AND.Q(I).GT.0)Q(I)=0.
        IF(N2(I).EQ.2.AND.Q(I).EQ.0)GO TO 210
    200 CONTINUE
        GO TO 208
    210 CONTINUE
        WRITE(6,6)
      6 FORMAT(/,3X,'FLOW RATE IS ZERO THEREFORE DESIRED FLCWING ',
       *' B.H.P. CAN NOT BE CALCULATED')
    208 CONTINUE
        IF(DI2.GT.0.)GO TC 204
        IF(DI.GT.4.227)FR=.10377/(DI**2.582)
        IF(DI.LF.4.227)FR=.10797/(DI**2.612)
        GO TO 207
    204 CONTINUE
        WRITE(6,11)
     11 FORMAT(//,3X,'ANNULAR FLOW CALCULATIONS WILL BE MADE IF TUEULAR',
       *' FLOW IS DESIRED SET DI2 = 0.0')
        D5=(((DI2-DI)**1.612)*(DI2+DI))**(1./2.612)
        IF(D5.LT.4.227)FR=.10797/(D5**2.612)
        IF(D5.GE.4.227)D6=(((DI2-DI)**1.582)*(DI2+DI))**(1./2.582)
        IF(D5.GE.4.227)FR=.10377/(D6**2.582)
    207 CONTINUE
        WRITE(6,50)
     50 FORMAT(//,3X,'BOTTCM-HOLE PRESSURE CALCULATIONS')
        WRITE(6,51)(A(I),I=1,20)
     51 FORMAT(/,3X,20A4)
        WRITE(6,52)
     52 FORMAT(/,3X,'W.H PRESSURE',3X,'CALC. BH PRESSURE',3X,'FLOW RATE')
        WRITE(6,53)
     53 FORMAT(7X,'PSIA',13X,'PSIA',13X,'MCF')
        DO 300 I=1,N1
        CALL BHPRS(PWF(I),TF(I),Q(I),DE,TRA,GG,C1,C2,C3,PC,TC,FR,N2(I)
       *,PBH)
        PBHF(I)=PBH
        WRITE(6,54)PWF(I),PBHF(I),Q(I)
     54 FORMAT(/,3X,F9.0,3X,F14.0,3X,F14.0)
    300 CONTINUE
        END
        SUBROUTINE PHPRS(P1,T3,Q,DE,TRA,GG,C1,C2,C3,PC,TC,FR,N2,PEH)
        REAL K2,K3,L,I1,M1,I2,N1,M2,M3,I3,N2,M4
   1000 CONTINUE
        IF(N2.EQ.1)S2=0.
        IF(N2.EQ.2)S2=1.
```

```
        TR=TRA
        IFLAG=1
        Z4=0.
        U3=TR

        U1=T3
        U2=(T3+TR)*.5
        T4=TR
        K2=18.75*GG*DE
        K3=2.*K2
        S3=1.
        F1=P1
        IFLAG=1
        GOTO 635
 325    P4=F7
        IF(S2.EQ.0..OR.S2.EQ.-1.)GOTO 375
        S3=-1
        F1=P4
        S2=0
        IFLAG=2
        GCTO 635
 360    S2=1
        P6=F7
        GOTO 396
 375    P6=P1
        CALL COMFAC(P4,TR,PC,TC,Z)
        Z4=Z
 396    GO TO 290
 635    IF(S3.GT.0.)GOTO 655
        Y1=U3
        Y3=U1
        GCTO 665
 655    Y1=U1
        Y3=U3
 665    CONTINUE
        CALL COMFAC(F1,Y1,PC,TC,Z)
        Z1=Z
        A1=F1/(Y1*Z1)
        D0=((FR*Q*.001)**2)
        I1=A1/(((A1**2)*.001)+S2*D0)
        M1=K2/(2*I1)
        DO 705 IXX=1,101
        F2=F1+S3*M1
        CALL COMFAC(F2,U2,PC,TC,Z)
        Z2=Z
        A2=F2/(U2*Z2)
        I2=A2/(((A2**2)*.001)+S2*D0)
        N1=I1+I2
        M2=K2/N1
        S1=M2*N1
        F4=F1+S3*M2
        IF(F4.EQ.F2)GOTO 780
        M1=M2
 705    CONTINUE
 780    M3=(K3-S1)/N1
        DO 785 IYY=1,101
        F3=F2+S3*M3
        CALL COMFAC(F3,Y3,PC,TC,Z)
        Z3=Z
        A3=F3/(Y3*Z3)
        I3=A3/(((A3**2)*.001)+S2*D0)
        N2=I2+I3
        M4=(K3-S1)/N2
        F5=F2+S3*M4
        IF(F5.EQ.F3)GOTO 855
        M3=M4
 785    CONTINUE
 855    D1=(3*K3)/(I1+4*I2+I3)
        F7=F1+S3*D1
        GOTO(325,360),IFLAG
 290    CONTINUE
        PBH=P4
        RETURN
        END
        SUBROUTINE COMFAC(XR,TR,PC,TC,Z)
 157    RRT=TC/TR
 158    RP=XR/PC
        A=0.06125*RPT*EXP(-1.2*(1.-RRT)**2)
        B=RRT*(14.76-9.76*RRT+4.58*RRT**2)
        C=RRT*(90.7-242.2*RRT+42.4*RRT**2)


        D=2.18+2.82*RRT
        Y=0.01
        DO 2 J=1,30
        IF(Y.GT.1.)Y=0.6
        IF(Y.LE.0.)Y=1.0E-5
```

```
166    F=-A*RP+Y*(1.+(Y*(1.+Y*(1.-Y))))/(1.-Y)**3-B*Y**2+C*Y**D
       IF(ABS(F)-1.0E-6)4,4,3
   3   DFDY=(1.+4.*Y*(1.+Y*(1.-Y))+Y**4)/(1.-Y)**4-2.*B*Y+D*C*Y**(D-1.)
   2   Y=Y-F/DFDY
   4   Z=A*RP/Y
       RETURN
1000   CONTINUE
       END
//GO.SYSIN DD *
PROBLEM NO. 12A      GREEN FIELD  WELL #1   P 238    SALT LAKE CITY    UTAH
   6  12.250. 8226.   2.411 0.0
   662. 398.    .0    .0    .0    .761
  2272. 78.0        4803.2
  2260. 78.0        5000.2
  2000. 78.0        5500.2
  2954. 60.0           0.1
  2900. 60.0           0.1
  2850. 60.0           0.1
/*
..
```

OUTPUT LISTING

ANNULAR FLOW CALCULATIONS WILL BE MADE IF TUBULAR FLOW IS DESIRED SET DI2 = 0.0

BOTTOM-HOLE PRESSURE CALCULATIONS

PROBLEM NO. 12A      GREEN FIELD  WELL #1   P 238     SALT LAKE CITY     UTAH

| W.H PRESSURE PSIA | CALC. BH PRESSURE PSIA | FLOW RATE MCF |
|---|---|---|
| 2284. | 2898. | 4803. |
| 2272. | 2883. | 5000. |
| 2012. | 2555. | 5500. |
| 2966. | 3750. | 0. |
| 2912. | 3686. | 0. |
| 2862. | 3626. | 0. |

**BIBLIOGRAPHY**

Interstate Oil Compact Commission: *Manual of Back-Pressure Testing of Gas Wells*, Oklahoma City (1972).

Smith, R. V., and Cullender, M. H.: "Practical Solution of Gas-Flow Equations for Wells and Pipelines with Large Temperature Gradients," *Trans.*, AIME (1956) 281–287.

**PURPOSE**

The program is designed to calculate stabilized absolute open flow potential (AOFP) for a gas well, utilizing modified isochronal test data in conjunction with extended flow (short time variation) data. The radius of drainage for the gas well should either be known or assumed (half the distance between the wells). This calculated stabilized AOFP is quite a reliable estimate and should be used for a low-permeability (relatively tight) reservoir. Stabilization, in this context, is referred to as the condition when the pressure transient or disturbance has reached the outer drainage boundary. Computation and theoretical considerations are based on Poettmann and Schilson's (1959) technical paper.

**METHOD**

$$Q_S = C_S(P_R^2 - P_f^2)^n$$

$$Q_{\text{SAOF}} = C_S P_R^{2n}$$

$$C_S = C_1\left(\frac{A + 1/2 \ln t_1}{A + 1/2 \ln t_s}\right)$$

$$A = \frac{C_2^{1/n}}{2(C_1^{1/n} - C_2^{1/n})} \ln t_2 - \frac{C_1^{1/n}}{2(C_1^{1/n} - C_2^{1/n})} \ln t_1$$

$$t_s = \frac{50\phi\mu C\gamma_e^2}{k}$$

where

$Q_S$ = Stabilized flow rate, MCFD;
$C_S$ = Stabilized performance coefficient;
$P_R$ = Reservoir pressure, psi;
$P_F$ = Flowing bottomhole pressure, psi;
$n$ = Performance exponent;
$Q_{SAOF}$ = Stabilized absolute open flow, MCFD;
$C_1$ = Performance coefficient for flow duration of $t_1$, hr;
$t_1$ = Isochronal flow period, hr;
$t_s$ = Stabilization time, day;
$C_2$ = Performance coefficient for flow duration of $t_2$, hr (extended flow time);
$t_2$ = Extended flow period, hr;
$\phi$ = Porosity, fraction;
$\mu$ = Gas viscosity, cp;
$C$ = Gas compressibility, vol/vol/psi;
$\gamma_e$ = Radius of drainage, ft;
$k$ = Reservoir permeability, md.

The modified isochronal test is conducted with the shut-in period (after each flow rate), about equal to the flow period. The associated unstabilized shut-in pressure is used for shut-in pressure ($P_{ts}$) in calculating the difference of pressures squared for the next flow period. The modified method does not yield a true isochronal curve but is completely satisfactory for determining the exponent ($n$) of the flow equation. To determine the position of the stabilized back-pressure curve, the well is produced at one rate for an extended period of time. Satisfactory results may be obtained with the modified isochronal test for determining the exponent ($n$) with flow test period of 1 to 2 hr. However, a much longer time period may be required for an extremely tight reservoir. Most of these considerations are applicable for radial flow geometry and are not applicable for hydraulically fractured tight gas wells, since the flow geometry (initially) is linear for such wells.

## PROGRAM DESCRIPTION

The program consists of a FORTRAN main program and two subprograms. Subprogram COMFAC is used to determine gas compressibility factors, which in turn yield gas compressibility values. Subprogram VISCO is used to estimate reservoir gas viscosity. Most of the computations are performed in the main program. Isochronal test data are used to determine the slope (exponent) of the performance curve and the coefficient $C$ by assuming a least-squares straight line fit through these points on a log-log basis. Restrictions are placed on the computed slope, such that $0.5 \leq n \leq 1$. Since these values are determined by least-squares fit of data, results will be obtained even with a set of bad data. Therefore, caution must be exercised and the data set should be checked by actual data point plot. Four isochronal data points and one extended time flow rate and pressure should be provided.

**Input Format**

| Card | Format | Content |
|---|---|---|
| 1 | 20A4 | A(I) |
| 2 | 20A4 | B(I) |
| 3 | F5.0, F4.0, F4.2, 3F4.3 | PR, TR, GG, CO2, XN2, HS2 |
| 4 | 2F4.3, 4F5.0, F3.1 | POR, GSAT, RD, PC, TC, PERM, XFT |
| 5–8 | 2F5.0, F8.0 | PSI(I), PFI(I), Q(I) |
| 9 | 2F5.0, F10.0, F5.1 | PSIE, PSFE, QE, TIE |

**Description of Input Parameters**

A(I) = Uses up to 80 alphanumeric characters for project identification;
B(I) = Uses up to 80 alphanumeric characters for project identification;
PR = Initial (static) reservoir pressure, psia;
TR = Reservoir temperature, °F;
GG = Gas gravity (air = 1);
CO2 = Mol fraction of carbon dioxide in the gas stream;
XN2 = Mol fraction of nitrogen in the gas stream;
H2S = Mol fraction of hydrogen sulfide in the gas stream;
POR = Average reservoir porosity, fraction;
GSAT = Gas saturation, fraction;
RD = Drainage radius, ft;
PC = Critical pressure for the gas, psia;
TC = Critical temperature for the gas, °R;
PERM = Average reservoir permeability, md;
XFT = Modified isochronal test flow/shut-in period;
PSI(I) = Bottomhole shut-in pressure prior to flow corresponding to each flow period, psi;
PFI(I) = Flowing bottomhole pressure for each flow period, psi;
Q(I) = Flow rate, each flow period, MCFD;
PSIE = Shut-in bottomhole pressure, corresponding to extended flow, psi;
PSFE = Flowing bottomhole pressure, extended flow period, psi;
QE = Flow rate (extended flow period), MCFD;
TIE = Time of extended flow, hr.

**EXAMPLE PROBLEM**

An isochronal (modified) flow test was performed on a gas well using an iso-chronal period of 1 hr and an extended flow period of 24 hr. The test results are shown in table 20.1. Other available information is as follows:

Average reservoir porosity = 20%,
Average gas saturation = 80%,
Average reservoir permeability = 15 md,
Average reservoir temperature = 120° F,
Initial reservoir pressure = 2905 psia,
Gas gravity = 0.60,
Gas critical pressure = 622 psia,
Gas critical temperature = 320° R,
Drainage radius = 1489 ft.

Determine stabilized absolute open flow potential of the well.

**Table 20.1 Well Test Data for Example Problem on Stablilized Absolute Open Flow Potential of a Gas Well**

| Time (hr) | Flow/Shut-in Period (hr) | Bottomhole Shut-in Pressure (psia) | Bottomhole Flowing Pressure (psia) | Flow Rate (MCFD) |
|---|---|---|---|---|
| 0 (1 A.M.) | Long shut-in to reservoir stabilization (opened at 1 A.M.) | 2905 | — | 0 |
| 1 (2 A.M.) | 1 (Flow) | — | 2817 | 1815 |
| 2 (3 A.M.) | 1 (Shut-in) | 2900 | — | 0 |
| 3 (4 A.M.) | 1 (Flow) | — | 2761 | 2725 |
| 4 (5 A.M.) | 1 (Shut-in) | 2895 | — | 0 |
| 5 (6 A.M.) | 1 (Flow) | — | 2707 | 3612 |
| 6 (7 A.M.) | 1 (Shut-in) | 2890 | — | 0 |
| 7 (8 A.M.) | 1 (Flow) | — | 2647 | 4525 |
| 30 (7 A.M.) | 24 (Flow) | — | 2300 | 4500 |

INPUT LISTING

FORTRAN CODING FORM

UU/CC   NAME
        DATE

PROGRAM
ROUTINE

0 = ZERO    1 = ONE      2 = TWO
Ø = ALPHA O  I = ALPHA I   Z = ALPHA Z

| STATEMENT NUMBER 1,2,3,4,5 | 6 | 7,8,9,10 ... 60 |
|---|---|---|
| 1. | | EXAMPLE PROBLEM 13   MØRGAN TIØNN   WESTVIRGINIA |
| 2. | | TEST 11/9,81   ANALYZED  11/12/1981 |
| 3. | | 2,9,0,5.1  20.0.60.00.00.0000 |
| 4. | | 2,0,0.8  001489.  622.  32.0.  15.11.0 |
| 5.1 | | 2905.  2817.  1815. |
| 5.2 | | 2900.  2761.  2725. |
| 5.3 | | 2895.  2707.  3612. |
| 5.4 | | 2890.  2647.  4525. |
| 6 | | 2890.  2300.  45001.  24. |

## SOURCE LISTING

```
C        THIS PROGRAM CALCULATES STABILIZED ABSOLUTE OPEN FLOW POTENTIAL
C        FROM FIVE POINT MODIFIED ISOCHRONAL TEST OF A GAS WELL.  THE
C        EFFECTIVE FORMATION PERMEABILITY IS DETERMINED SEPARATELY FROM
C        BUILD-UP DATA AND USED IN THIS PROGRAM.  THIS COMPUTATION IS
C        VALID WITH RADIAL FLOW ASSUMPTIONS.
         DIMENSION A(20),B(20),PSI(20),PFI(20),Q(20),X(20),Y(20),XY(20),
        *X2(20),Y2(20)
 1000 CONTINUE
         WRITE(6,1)
    1 FORMAT(/,3X,'DETERMINATION OF STABILIZED ABSOLUTE OPEN FLOW',
        *' POTENTIAL FOR GAS WELL FROM MODIFIED ISOCHRONAL DATA')
         READ(5,2)(A(I),I=1,20)
         READ(5,2)(B(I),I=1,20)
         WRITE(6,3)(A(I),I=1,20)
         WRITE(6,3)(B(I),I=1,20)
    2 FORMAT(20A4)
    3 FORMAT(//,3X,20A4)
         SUMX=0.
         SUMY=0.
         SUMXY=0.
         SUMX2=0.
         SUMY2=0.
         READ(5,5)PR,TR,GG,CO2,XN2,H2S
    5 FORMAT(F5.0,F4.0,F4.2,3F4.3)
         READ(5,6)POR,GSAT,RD,PC,TC,PERM,XFT
    6 FORMAT(2F4.3,4F5.0,F3.1,F5.1)
C        PR=RESERVOIR PRESSURE, PSI
C        TR=RESERVOIR TEMP, FAHENHEIT
C        GG=GAS GRAVITY (AIR=1.)
C        CO2=MOLE FRACTION CO2 IN THE AIR
C        XN2=MOL FRACTION N2 IN THE GAS
C        H2S=MOL FRACTION H2S IN THE GAS
C        POR=EFFECTIVE POROSITY FRACTION
C        GSAT=GAS SATURATION (1.-WATER SATURATION), FRACTION
C        RD=RADIUS OF DRAINAGE, FT
C        PC=CRITICAL PRESSURE, PSIA
C        TC=CRITICAL TEMPERATURE, RANKINE
C        PERM=EFFECTIVE PERMEABILITY TO GAS, MD

C        XFT=ISOCHRONAL FLOW PERIOD, HR
         DO 100 I=1,4
         READ(5,4)PSI(I),PFI(I),Q(I)
         X(I)=ALOG10(PSI(I)**2-PFI(I)**2)
         Y(I)=ALOG10(Q(I))
         X2(I)=X(I)**2
         Y2(I)=Y(I)**2
         XY(I)=X(I)*Y(I)
         SUMX=SUMX+X(I)
         SUMY=SUMY+Y(I)
         SUMXY=SUMXY+XY(I)
         SUMX2=SUMX2+X2(I)
         SUMY2=SUMY2+Y2(I)
    4 FORMAT(2F5.0,F8.0)
  100 CONTINUE
         SLOPE=(SUMXY-SUMX*SUMY/4.)/(SUMX2-SUMX**2/4.)
         B1=.25*SUMY-SLOPE*.25*SUMX
         C1=10.**B1
         IF(SLOPE.GT.1.)SLOPE=1.
         IF(SLOPE.LT.0.5)SLOPE=.5
         XN1=SLOPE
         READ(5,7)PSIE,PSFE,QE,TIE
    7 FORMAT(2F5.0,F10.0,F5.1)
C        PSIE=SHUT-IN PRESSURE BEFORE EXTENDED FLOW, PSI
C        PSFE=FLOWING PRESSURE AT THE END OF EXTENDED FLOW, PSI
C        TIE=EXTENDED FLOW TIME, HRS.
C        QE=RATE AT THE END OF EXTENDED FLOW, MCFD
C        JJ=ISOCHRONAL FLOW POINT 1,2,3, OR 4 TO BE SELECTED FOR COMPUTATION
  102    C2=QE/((PSIE**2-PSFE**2)**XN1)
         IS=1./SLOPE
         XX=C1**IS-C2**IS
         XXX=((.5*C2**IS)/XX)*ALOG(TIE)-((.5*C1**IS)/XX)*ALOG(XFT)
         CALL VISCO(PR,TR,PC,TC,GG,CO2,XN2,H2S,VIS)
         U1=VIS
  106    RP=PR/PC
  107    RT=(TR+460.)/TC
         CALL COMFAC(RP,TR,PC,TC,Z)
         E1=RP
         W=Z
         F1=E1-.04
```

```
        RP=F1
        CALL COMFAC(RP,TR,PC,TC,Z)
        W1=Z
        F2=E1-.02
        RP=F2
        CALL COMFAC(RP,TR,PC,TC,Z)
        W2=Z
        F3=E1+.02
        RP=F3
        CALL COMFAC(RP,TR,PC,TC,Z)
        W3=Z
        F4=E1+.04
        RP=F4
        CALL COMFAC(RP,TR,PC,TC,Z)
        W4=Z
        E5=F1+F2+F3+F4
        Y8=W1+W2+W3+W4
        E6=F1*W1+F2*W2+F3*W3+F4*W4
        D5=F1**2+F2**2+F3**2+F4**2
108     XK5=((Y8*E5)-(4.*E6))/((E5**2)-(4.*D5))
109     XL5=(1./E1)-(1./W)*XK5
110     XM5=XL5/PC
111     SS=SQRT((.00088*PERM)/(U1*POR*GSAT*XM5))
112     TS=(RD/SS)**2
113     C6=C1*((XXX+.5*ALOG(XFT))/(XXX+.5*ALOG(TS)))**SLOPE
        SOFP=C6*PR**(2.*XN1)
        WRITE(6,10)SOFP
   10   FORMAT(//,3X,'STABILIZED AOF',3X,F10.2,' MCFD')
        WRITE(6,11)TS
   11   FORMAT(/,3X,'STABILIZATION TIME',3X,F10.2,' HRS')
        WRITE(6,12)XN1
   12   FORMAT(/,3X,'SLOPE OF THE FLOW CURVE',3X,F3.2)
 1001   CONTINUE
        STOP
        END
        SUBROUTINE COMFAC(T,S,PC,TC,Z)
        RP=T
        X=S+460.
        RRT=TC/X
        A=0.06125*RRT*EXP(-1.2*(1.-RRT)**2)
        B=RRT*(14.76-9.76*RRT+4.58*RRT**2)
        C=RRT*(90.7-242.2*RRT+42.4*RRT**2)
        D=2.18+2.82*RRT
        Y=0.01
        DO 2 J=1,30
        IF(Y.GT.1.)Y=0.6
   10   F=-A*RP+Y*(1.+(Y*(1.+Y*(1.-Y))))/(1.-Y)**3-B*Y**2+C*Y**D
        IF(ABS(F)-1.0E-6)4,4,3
    3   DFDY=(1.+4.*Y*(1.+Y*(1.-Y))+Y**4)/(1.-Y)**4-2.*B*Y+D*C*Y**(D-1.)
    2   Y=Y-F/DFDY
    4   Z=A*RP/Y
        RETURN
        END
        SUBROUTINE VISCO(PB,TRE,PC,TC,GG,C1,C2,C3,VIS)
        REAL M
        DATA B0,B1,B2,B3,B4,B5,B6,B7,B8/1.112391E-2,1.677266E-5,2.113605E
       *-9,-1.094850E-4,-6.403164E-8,8.993745E-11,4.577352E-7,2.129034E-1
       *0,3.977322E-13/
        DATA A0,A1,A2,A3,A4,A5,A6,A7,A8,A9/-2.462118,2.970547,-2.862641E-1
       *,8.054205E-3,2.808609,-3.498033,3.603730E-1,-1.044324E-2,-7.93385
       *7E-1,1.396433/
        DATA H0,H1,H2,H3,H4,H5/-1.491449E-1,4.410155E-3,8.393872E-2,-1.86
       *4088E-1,2.033679E-2,-6.095793E-4/
        TR=TRE+460.
        RP=PB/PC
        RT=TR/TC
        IF(RT.GT.3.)GO TO 20
        IF(RT.LT.1.2)GO TO 30
        IF(RP.GT.20.) GO TO 40
        M=28.97*GG
        Y5=GG/(39.8782+64.0537*GG)
        Y6=1.03401E-2-3.4823E-3/GG
        Y7=8.08098E-3-3.99988E-3/GG
        U=TR-460.
        V1=C1*Y5+C2*Y6+C3*Y7
        V2=B0+B1*U+B2*U**2+B3*M+B4*U*M+B5*U**2*M+B6*M**2+B7*U*M**2+B8*U**
       *2*M**2
        V3=((V1+V2)*1.E05+0.5)*1.E-5
        IF(RP.LT.1.)GO TO 50
        V4=A0+A1*RP+A2*RP**2+A3*RP**3+RT*(A4+A5*RP+A6*RP**2+A7*RP**3)
        V5=RT**2*(A8+A9*RP+H0*RP**2+H1*RP**3)
        V6=RT**3*(H2+H3*RP+H4*RP**2+H5*RP**3)
        V7=V4+V5+V6
        V8=EXP(V7)/RT
        U1=((V3*V8)*1.E4+0.5)*1.E-4
        GO TO 65
   20   WRITE(6,21)
```

```
      21 FORMAT(//,3X,'PSEUDO REDUCED TEMPERATURE IS GREATER THAN 3.0')
      30 WRITE(6,31)
      31 FORMAT(//,3X,'PSEUDO REDUCED TEMPERATURE IS LESS THAN 1.2')
      40 WRITE(6,41)
      41 FORMAT(//,3X,'PSEUDO REDUCED PRESSURE IS GREATER THAN 20')
         GO TO 60
      50 U1=V3
         GO TO 65
      60 U1=0.
      65 CONTINUE
         VIS=U1
         RETURN
         END
//GO.SYSIN DD *
EXAMPLE PROBLEM 13            MORGANTOWN      WEST VIRGINIA
TEST           1/1/1981       ANALYZED 1/12/81
2000.120.  .60.000.000.000
.200.8001489. 622. 320.  15.1.
2905.2817.   1815.
2900.2761.   2725.
2895.2707.   3612.
2890.2647.   4525.
2890.2300.   4500.   24.0
```

# OUTPUT LISTING

DETERMINATION OF STABILIZED ABSOLUTE OPEN FLOW POTENTIAL FOR GAS WELL FROM MODIFIED ISOCHRONAL DATA

EXAMPLE PROBLEM 13          MORGANTOWN          WEST VIRGINIA

TEST          1/1/1981          ANALYZED 1/12/81

STABILIZED AOF          4577.79 MCFD

STABILIZATION TIME          199.78 HRS

SLOPE OF THE FLOW CURVE          .93

**BIBLIOGRAPHY**

Guerrero, E. T.: *Practical Reservoir Engineering*, The Petroleum Publishing Company, Tulsa (1968).

Katz, D. L., Cornell, D., Kobayashi, R., Poettman, F. H., Vary, J. A., Elenbaas, J. R., and Weinaug, C. F.: *Handbook of Natural Gas Engineering*, McGraw-Hill, New York (1959).

Poettmann, F. H., and Schilson, R. E.: "Calculation of the Stabilized Performance Coefficient of Low Permeability Natural Gas Wells," *Trans.*, AIME (1959) 240–246.

# 21 CONVERSION OF POINT-AFTER-POINT GAS WELL TEST RESULTS TO EQUIVALENT ISOCHRONAL TEST RESULTS

## PURPOSE

A conventional back-pressure test, or point-after-point test, yields a back-pressure curve with an incorrect slope because the drainage area differs for various flow rates and the log-log plot of $\Delta(p2)$ versus $Q$ tends to shift to the left.

Isochronal test data are normally restricted to the same drainage area for all flow rates and thus provide a more accurate slope than the conventional test data. Therefore, in certain situations it may be desirable to convert the conventional back-pressure curve to an equivalent isochronal back-pressure curve and thus to obtain a more accurate slope value.

This program is designed to perform the necessary computations for such conversions by utilizing the superimposition principle and a dimensionless pressure response function–dimensionless time relationship. The procedure is valid for dimensionless time values greater than 1000.

## METHOD

$$T_D = \frac{6.33 \times 10^{-3} k P_{avg} t}{\mu_{avg} \phi_{HC} \gamma_w^2}$$

$$P_T = \frac{1}{2} (ln\ T_D + 0.80907)$$

$$(P_S^2 - P^2)_{isochronal,\ j} = (P_S^2 - P^2)_{conventional,\ j} \times \frac{Q_j^{1/n}\ P_T^{1/n}}{\displaystyle\sum_{i=1}^{j} (Q_i^{1/n} - Q_{(i-1)}^{1/n})\ P_{T(j+1-i)}^{1/n}}$$

where

$T_D$ = Dimensionless time;

$k$ = Effective permeability to gas;

$P_{avg}$ = Average reservoir flowing pressure, psi;

$t$ = Time in days;

$\mu_{avg}$ = Gas viscosity at average reservoir flow condition, cp;

$\phi_{HC}$ = Hydrocarbon porosity, fraction;

$\gamma_W$ = Effective well radius, ft;

$P_T$ = Dimensionless cumulative pressure drop function;

$P_T = f(T_D)$;

$Q$ = Gas flow rate, ·MMSCFD;

$P_s$ = Static (shut-in) reservoir pressure, psia;

$P$ = Flowing bottomhole pressure, psia;

$j$ = Subscript denoting data for selective cumulative time.

## PROGRAM DESCRIPTION

The program consists of a simple FORTRAN main program that computes the equivalent isochronal difference in the squared pressure values for each conventional flow point. Using the flow rates and respective differences in squared pressure (equivalent isochronal), a log-log least-squares line through these data points is hypothesized. The slope of this hypothetical straight line is assumed to be the equivalent isochronal slope.

### Input Format

| Card | Format | Content |
|------|--------|---------|
| 1 | 20A4 | A(J) |
| 2 | I2 | N |
| 3 | F5.3, F5.0, 2F5.3, 2F5.0, F5.3 | POR, PERM, SW, VIS, RE, RP, SLOPE |
| 4–N | 2F5.0, F10.0 | T(J), BHP(I), FR(I) |

### Description of Input Parameters

A(I) = Uses up to 80 alphanumeric characters for project identification;

N = Number of data points given;

POR = Average porosity, fraction;

PERM = Effective gas permeability, md;

SW = Initial water saturation;

VIS = Average gas viscosity, cp;
RE = Effective well radius, ft;
RP = Shut-in reservoir pressure, psia;
SLOPE = Slope of the conventional back-pressure curve;
T(J) = Cumulative flowing time, hr;
BHP(I) = Bottomhole flowing pressure corresponding to time T(I), psi;
FR(I) = Flow rate corresponding to time T(I), MMSCF.

## EXAMPLE PROBLEM

A back-pressure test (conventional) was performed on a gas well, and the results are shown in table 21.1. Determine equivalent isochronal slope. Other available data are as follows:

Effective gas permeability = 80 md,
Viscosity of gas at reservoir condition = 0.027 cp,
Effective porosity = 28%,
Interstitial water saturation = 42.9%,
Effective well radius = 0.5 ft,
Reservoir static pressure = 2905 psig,
Slope of the conventional back-pressure curve = 0.892,
Barometric pressure = 14.7.

Table 21.1 Conventional Back-Pressure Test Data

| Time (hr) | Bottomhole Pressure (psi) | Flow Rate (MMSCF) |
|-----------|---------------------------|-------------------|
| 0 | 2905 | 0 |
| 1 | 2817 | 1.815 |
| 2 | 2761 | 2.725 |
| 3 | 2707 | 3.612 |
| 4 | 2647 | 4.525 |

**INPUT LISTING**

# FORTRAN CODING FORM

UU/CC  NAME ____  DATE ____

PROGRAM ____
ROUTINE ____

0 = ZERO   Ø = ALPHA O   1 = ONE   I = ALPHA I   2 = TWO   Z = ALPHA Z

| CARD | STATEMENT NUMBER 1,2,3,4,5 | 6 | 7,8,9,10,11,...,80 |
|---|---|---|---|
| 1. | B,A,N,I | G,H,Ø,S,H, #,1, ,S,A,V,G,A,S, ,C,A,L,L,I,F,Ø,R,N,I,A, ,P,R,Ø,B,L,E,M, ,1,4, ,P, ,2,5,1 |
| 2. | 4 | | |
| 3. | .,1,2,8,0 | 8,.,0, ,.,4,2,9, ,.,0,2,7, ,.,5,0,2,9,0,5,·,.,0,1,8,9,1,2 |
| 4-1 | 1,. | 2,8,.,1,7,. ,1,8,.,1,5,·, |
| 4-2 | 2,. | 2,7,6,. ,2,7,2,5,·, |
| 4-3 | 3,. | 2,7,0,7,. ,3,6,1,2,·, |
| 4-4 | 4,. | 2,6,7,. ,4,5,2,5,·, |

# SOURCE LISTING

```
      THIS PROGRAM CONVERT CCNVENTIONAL POINT AFTER POINT GAS WELL FLOW
      TEST DATA TO EQUIV. ISOCHRONAL FLOW TEST RESULT

      DIMENSION A(20),T(5),BHP(5),FR(5),TD(5),C2(5),C6(5),C7(5),C8(5),
     *C9(5),C10(5),C11(5),C12(5),C13(5)
 1000 CONTINUE
      READ(5,3)(A(I),I=1,20)
      N=NUMBER OF DATA POINTS OR MEASUREMENTS
      READ(5,2)N
    2 FORMAT(I2)
      READ(5,1)POR,PERM,SW,VIS,RW,RP,SLOPE
    1 FORMAT(F5.3,F5.0,2F5.3,F5.2,F5.0,F5.3)
      POR=EFFECTIVE PORISITY OF THE RESERVOIR (AVG.), FRACTION
      PERM=EFFECTIVE PERMEABILITY TO GAS, MD
      SW=WATER SATURATION, FRACTION
      VIS=GAS VISCOSITY AT RESERVOIR CONDITION, CP
      RW=EFFECTIVE WELL-BORE RADIUS, FT
      RE=WELL DRAINAGE RADIUS OR WELL SPACING (HALF THE DISTANCE
      BETWEEN THE WELL), FT
      RP=RESERVOIR PRESSURE, PSIA
      TDF=0.00633*PERM*RP/((1.-SW)*POR*VIS*RW**2)
      SLOPE=RECIPROCAL CF CONVENTIONAL SLOPE OF BACK-PRESSURE CURVE
    3 FORMAT(20A4)
      DO 10 I=1,N
      READ(5,4)T(I),BHP(I),FR(I)
      T(I)CUM FLOWING TIME FROM S. I. CONDITIONS, HRS
      BHP(I)=BOTTOM-HOLE FLOWING PRESSURE, PSIA
      FR(I)=FLOW RATE, MCFD
    4 FORMAT(2F5.0,F10.0)
   10 CONTINUE
      DO 15 I=1,N
      TD(I)=TDF*T(I)*.04167
   15 CONTINUE
      DO 20 I=1,N
      C2(I)=(RP**2.-BHP(I)**2)
      C6(I)=0.5*(ALOG(TD(I))+.80907)
      C7(I)=FR(I)**(1./SLOPE)
      IF(I.EQ.1)C8(1)=C7(1)
      IF(I.GT.1)C8(I)=C7(I)-C7(I-1)
      C9(I)=C6(I)**(1./SLOPE)
      C10(I)=C7(I)*C9(1)
   20 CONTINUE
      DO 30 I=1,4
      SUM=0.
      DC 35 J=1,I
      SUM=SUM+C8(J)*C9(I-J+1)
   35 CONTINUE
      C11(I)=SUM
   30 CONTINUE
      DO 36 I=1,N
      C12(I)=C10(I)/C11(I)
      C13(I)=C12(I)*C2(I)
   36 CONTINUE
      SUMX=0.
      SUMX2=0.
      SUMY=0.
      SUMY2=0.
      SUMXY=0.
      DO 40 I=1,N
      SUMX=SUMX+ALOG10(FR(I))
      SUMY=SUMY+ALOG10(C13(I))
      SUMX2=SUMX2+(ALOG10(FR(I)))**2
      SUMY2=SUMY2+(ALOG10(C13(I)))**2
      SUMXY=SUMXY+ALOG10(FR(I))*ALOG10(C13(I))

   40 CONTINUE

      SLOPIS=((SUMXY)-SUMX*SUMY/FLOAT(N))/((SUMX2)-SUMX**2/FLOAT(N))
      CEPT=(SUMY/FLOAT(N)-SLOPIS*SUMX/FLOAT(N))
      RESL=1./SLOPIS
      WRITE(6,50)
   50 FORMAT(//,3X,'CALCULATIONS FOR CONVERTING POINT AFTER POINT',
     *' BACK-PRESSURE DATA TO ISCHRONAL BACK-PRESS.')
      WRITE(6,51)(A(I),I=1,20)
   51 FORMAT(///,17X,20A4)
      WRITE(6,52)
   52 FORMAT(//,3X,'TIME',3X,'FLCW RATE',3X,'DIF.PR.SQ.',3X,'CONV.SLOPE'
     *,3X,'ISO.SLOPE',3X,'DIF.PR.SQ.ISO')
      WRITE(5,53)
   53 FORMAT(4X,'HR',7X,'MCFD',5X,'PSIA*PSIA',5X,'DIM.LESS',3X,
```

```
      *'DIM.LESS',3X,'PSIA*PSIA')
       DO 60 I=1,N
       WRITE(6,55)T(I),FR(I),C2(I),SLOPE,RESL,C13(I)
   55 FORMAT(F6.0,F11.0,F14.0,F12.3,F12.3,F15.0)
   60 CONTINUE
       GO TO 57
   25 WRITE(6,58)
   58 FORMAT(3X,'FLOW TEST DURATICN ARE TOO SMALL, USE LONGER DURATION',
      *' DATA IF AVAILABLE')
   57 CONTINUE
       STOP
       END
//GO.SYSIN DD *
BANI GHOSH #1      SAUGAS        CALIFORNIA      PROBLEM 14      P251
   4
  .280  80. .429 .027  .502905.0.892
    1.2817.     1815.
    2.2761.     2725.
    3.2707.     3612.
    4.2647.     4525.
/*
```

**OUTPUT LISTING**

CALCULATIONS FOR CONVERTING POINT AFTER POINT BACK-PRESSURE DATA TO ISCHRONAL BACK-PRESS.

|  | BANI GHOSH #1 | SAUGAS | CALIFORNIA | PROBLEM 14 | P251 |
|---|---|---|---|---|---|

| TIME | FLOW RATE | DIF.PR.SQ. | CONV.SLOPE | ISO.SLOPE | DIF.PR.SQ.ISO |
|---|---|---|---|---|---|
| 1. | 1815. | 503538. | 0.892 | 0.949 | 503538. |
| 2. | 2725. | 815906. | 0.892 | 0.949 | 782976. |
| 3. | 3612. | 1111178. | 0.892 | 0.949 | 1041993. |
| 4. | 4525. | 1432418. | 0.892 | 0.949 | 1321924. |

**BIBLIOGRAPHY**

Guerrero, E. T.: *Practical Reservoir Engineering,* The Petroleum Publishing Company, Tulsa (1968).

## PURPOSE

The program computes gas deliverability to the pipeline from a gas well using bottomhole conditions and the vertical-friction-type calculations (as described in the IOCC manual of back pressure testing of natural gas wells) in conjunction with the gas material balance relationship. Gas material balance may be based either on pressure or the pressure-over-compressibility-factor and cumulative production relationship.

## METHOD

$$Q = C(P^2 - P_f^2)^n,$$

$$P/Z = P_i/Z_i - mG_P,$$

or

$$P = P_i - mG_P,$$

where

$Q$ = Flow rate, MCFD;
$P$ = Static bottomhole/reservoir pressure, psi;
$P_f$ = Flowing bottomhole pressure, psi;
$P_i$ = Initial reservoir pressure, psi;
$Z$ = Gas compressibility factor;
$Z_i$ = Initial gas compressibility factor;
$G_P$ = Cumulative gas production, MMCF;
$n$ = Slope of the back-pressure curve.

The availability portion of the program has the flexibility to follow a fixed or variable contract obligation schedule on an annual basis. Either the deliverability pressure or the fixed annual volume obligation may be varied on a year-to-year basis. Volume obligations are provided in the form of a fixed annual quantity and a percentage of the maximum peak rate. All computations are made on an annual basis.

As a starting point, the program computes the maximum gas flow rate from a reservoir through a production string against a flowing wellhead pressure. This computation is carried out by an iterative procedure with rates based on percentages of the absolute open flow potential.

The flowing wellhead pressure is the lesser of the input wellhead flowing pressure or a predesignated fraction of the static wellhead pressure (DSP factor). The following wellhead pressure may thus be allowed to decline by proper use of the DSP factor to a pre-established minimum flowing wellhead pressure.

To flag possible liquid removal problems or as an aid to designing a proper production string, the maximum flow rates necessary to lift liquids up to a gas-liquid ratio of 130 bbl/MMCF are also computed.

## PROGRAM DESCRIPTION

The program consists of a FORTRAN main program and four subprograms—namely, BHFP, PRESS, COMFAC, and COMFC 1. Subprogram BHFP provides the bottomhole flowing pressure values for a given wellhead pressure and flow rate. Subprogram PRESS is utilized to compute bottomhole static pressure values from $P/Z$ values. COMFAC and COMFC 1 both provide the gas compressibility factor values. The main program is complex in structure, with several loops, and uses an iterative procedure for convergence at each computational step.

## Input Format

| Card | Format | Content |
|------|--------|---------|
| 1 | 20A4 | ANAME(I) |
| 2 | 20A4 | BNAME(I) |
| 3 | 5F10.3 | BO, XMFP, RAP, RES, L |
| 4 | 5F10.3 | H, TO, T3, T4, GG |
| 5 | 5F10.3 | C1, C2, C3, B5, T5 |
| 6 | 5F10.3 | D4, D3, RSIP, AOFP |
| 7 | 5F10.3 | NO, G1, Q2, ORF, DSP |
| 8 | 5F10.3 | XLY, XYS, L1, L5 |
| 9 | 5F10.3 | V1, P1, ORF |
| (Required only if L1 = 0) | | |

| 10 | 5F10.3 | Y6, V1, P1, ORF |

(Required only if
  L1 = 1)

| 11 | 5F10.3 | V1, P1, ORF |

(Required only if
  L1 = 1)

---

**Description of Input Parameters**

ANAME(I) = Uses up to 80 alphanumeric characters for project identification;
BNAME(I) = Uses up to 80 alphanumeric characters for project identification;
     BO = Barometric pressure, psia;
   XMFP = Minimum wellhead flowing pressure, psig;
    RAP = Reservoir abandonment pressure, psig;
    RES = Reserves, MMFC;
      L = Length of the flow string, ft;
      H = Depth midpoint perforation, ft;
     TO = Static wellhead temperature, °F;
     T3 = Flowing wellhead temperature, °F;
     T4 = Reservoir temperature, °F;
     GG = Gas gravity (air = 1);
     C1 = Mol fraction of carbon dioxide;
     C2 = Mol fraction of nitrogen;
     C3 = Mol fraction of hydrogen sulfide;
     B5 = Critical pressure, psia;
     T5 = Critical temperature, °R;
     D4 = Inside diameter of tubing or casing, in;
     D3 = Outside diameter of tubing (must be zero for tubular flow), in;
   RSIP = Static reservoir pressure, psia;
   AOFP = Stabilized absolute open flow potential, MCFD;
     NO = Reservoir back-pressure slope;
     G1 = Option flag: $G1 = 0$, $P$ versus $G_P$ equation is to be used;
          $G1 = 1$, $P/Z$ versus $G_P$ equation is to be used;
     Q2 = Abandonment rate, MCF;
    ORF = Obligation reduction factor based on gas sales contract
          (normally 0.75 to 0.8; however, can be considerably lower or
          higher but cannot exceed 1);
    DSP = Declining static pressure factor—that is, a factor indicating
          wellhead flowing pressure as a fraction of wellhead static
          pressure to be used for deliverability prediction;
    XLY = First year (e.g., 84) of the prediction;
    XYS = Years of prediction desired;
     L1 = Option flag: 0 = constant annual production or constant line-
          pressure-type gas contract schedule, 1 = variable annual
          production or wellhead flowing-pressure-type gas contract
          schedule;

L5 = Number of years annual production and/or line pressure is desired (i.e., input of years annual volume/wellhead pressures V1/P1 vary. After L5 years, input V1 and P1 once in card);

V1 = Contract annual volume desired, MMCF;

P1 = Initial wellhead flowing pressure, psig;

Y6 = Year.

*Note:* If L1 = 0, input card 9 once only. If L1 = 1, input card 10 L5 times— that is, for L5 years. After L5 years, input card 11 once.

### EXAMPLE PROBLEM

Prepare a 20-year gas deliverability forecast for a gas well with the following data. The gas sales contract calls for a maximum deliverability of 200, 175, and 150 MMCF for the first 3 years respectively or at 100% of maximum capacity at 800-psig wellhead flowing pressure. Minimum wellhead pressure is 800 psig. After 3 years, maximum deliverability is 150 MMCF/yr, or 75% of maximum capacity.

Barometric pressure = 12 psia,
Minimum wellhead flowing pressure = 800 psig,
Reservoir abandonment pressure = 1000 psig,
Reserves = 8000 MMCF,
Length of the flow string = 8000 ft,
Depth of the midpoint perforation = 7050 ft,
Static wellhead temperature = 60° F,
Flowing wellhead temperature = 80° F,
Reservoir temperature = 220° F,
Gas gravity = 0.65 (air = 1),
Mol fraction of carbon dioxide in the gas = 0,
Mol fraction of nitrogen in the gas = 0,
Mol fraction of hydrogen sulfide in the gas = 0,
Critical pressure of the gas = not known,
Critical temperature of the gas = not known,
I.D. of tube = 2.41 in,
O.D. of tube = 2.58 in,
I.D. of casing = 4.5 in,
Type of flow = tubular,
Static reservoir pressure = 3450 psia,
Absolute open flow potential (stabilized) = 3000 MCFD,
Reservoir back-pressure slope = 0.75,
Abandonment rate = 10 MCFD,
Obligation reduction factor = 0.75,
DSP factor = 0.99,
First year of prediction = 1971,
Number of years projection required = 20 years.

**INPUT LISTING**

FORTRAN CODING FORM

UU/CC   NAME
        DATE

0 = ZERO      1 = ONE       2 = TWO
Ø = ALPHA O   I = ALPHA I   Z = ALPHA Z

PROGRAM
ROUTINE

| STATEMENT NUMBER | | | | | | | | | | | | | |
|---|---|---|---|---|---|---|---|---|---|---|---|---|---|

```
A N U R A D H A   S I N H A
H Ø L L A D A Y             S A L T L A K E  C I T Y      U T A H
        1 2 .        8 0 0 .        1 0 0 0 .        8 0 0 0 .        8 0 0 0 .
7 0 5 0 .            6 0 .          8 0 .            2 0 0 .          6 5 .
0 .        1 0 . 0        1 0 . 0        1 0 .            0 .        1 0 .
2 . 4 1        0 . 0        3 4 5 0 .        3 0 0 0 .        9 9 1
      . 7 5 .            1 . 0            1 0 .            7 5 .
7 1 .              2 0 .            1 .              3 .
7 1 .              2 0 0 .          8 0 0 .          1 .
7 2 .              1 7 5 .          8 0 0 .          1 .
7 3 .              1 5 0 .          8 0 0 .          1 .
1 5 0 .            8 0 0 .          1 7 5 .
```

G R E E N F I E L D          R A T H M Ø R E  E S T A T E

## SOURCE LISTING

```
C     THIS PROGRAM COMPUTES GAS WELL DELIVERABILITY/PRODUCTION FORECAST
C     BY UTILIZING GAS FLOW EQUATIONS AND STEP-WISE PROCEDURE IN IOCC
C     MANUAL
      DIMENSION V(4,4),B(4,4),E(4,4),A(4,4),ANAME(20),BNAME(20)
      DIMENSION C(20,3),S(6),R(15,3)
      REAL I3,K0,K6,L2,M2,N1,I5,K2,K8,L4,M3,N2,I6,K3,K9,L5,M4,I1,I7,K4,
     *L,M,M5,I2,I8,K5,L1,M1,N0
      READ(5,100)(ANAME(I),I=1,20)
      READ(5,100)(BNAME(I),I=1,20)
100   FORMAT(20A4)
      WRITE(6,101)(ANAME(I),I=1,20)
      WRITE(6,101)(BNAME(I),I=1,20)
101   FORMAT(3X,20A4)
      READ(5,102)B0,XMFP,RAP,RES,L
      READ(5,102)H,T0,T3,T4,G
      READ(5,102)C1,C2,C3,B5,T5
      READ(5,102)D4,D3,RSIP,AOFP
      READ(5,102)N0,G1,Q2,ORF,DSP
102   FORMAT(5F10.3)
      U7=0.365
      D5=(((D4-D3)**1.612)*(D4+D3))**(1./2.612)
      IF(D5.LT.4.277)D=0.10915/(D5**2.612)
      IF(D5.GE.4.277)D6=(((D4-D3)**1.582)*(D4+D3))**(1./2.582)
      IF(D5.GE.4.277)D=0.10450/(D6**2.582)
      CI=AOFP/(RSIP**(2*N0))

      K2=18.75*G*H
      K3=2.*K2
      U0=T0+460.
      U1=T3+460.
      U3=T4+460.
      U2=0.5*(U1+U3)
      U4=0.5*(U0+U3)
      IF(B5.LE.0.0)B5=676.2366+6.9868*G-25.8273*G*G+432.9*C1-167.3*C2+65
     *4.*C3
      IF(T5.LE.0.0)T5=142.7712+380.9671*G-36.405*G*G-167.3*C1-279.9*C2+
     *127.*C3
220   A4=RAP+B0
      IF(G1.EQ.0.) GO TO 295
      CALL COMFAC(RSIP,U3,B5,T5,Z)
      Z5=Z
      IF(A4.EQ.0.) GO TO 275
      CALL COMFAC(A4,U3,B5,T5,Z)
      Z4=Z
      GO TO 285
275   K4=(RSIP/Z5)/RES
      GO TO 300
285   K4=((RSIP/Z5)-(A4/Z4))/RES
      GO TO 300
295   K4=(RSIP-A4)/RES
300   WRITE(6,15)
      WRITE(6,480)B0,XMFP,RAP,RES
      WRITE(6,500)L,T3,H,T0,G,C1,C2,C3,B5,T5
15    FORMAT(/10X,'PROGRAM : GAS DELIVERABILITY FORECAST'/)
      IF(G1.EQ.0.) GO TO 415
      WRITE(6,535)K4
      GO TO 420
415   WRITE(6,535)K4
420   WRITE(6,540)DSP,T4,N0,ORF
440   WRITE(6,550)
475   CONTINUE
595   READ(5,102,END=9999)XLY,XYS,L1,L5
      ILY=XLY
      IYS=XYS
      IF(L1.GT.0.) GO TO 610
605   READ(5,102,END=9999)V1,P1,ORF
610   IX4=0.
      Y0=0.
611   IX4=IX4+1
      Y0=Y0+1
      IF(L1.EQ.0.)GO TO 660
      IF(L5+1..EQ.Y0) GO TO 640
      READ(5,102,END=9999)Y6,V1,P1,ORF
      GO TO 655
640   READ(5,102,END=9999)V1,P1,ORF
      L1=0.
      GO TO 660
655   Q3=V1/U7
```

```
660 IF(P1.EQ.XMFP)GO TO 730
    S3=-1.
    S2=1.
    Q=0.
    F1=RSIP
    CALL  BHFP(U0,U1,U2,U3,U4,F1,B5,T5,L,H,D,Q,S2,S3,K2,K3,IFLAG,F7)
    IF(IFLAG.GT.1)GO TO 710
690 F0=F7-B0
    IF(XMFP.GT.F0)GO TO 2120
    G7=F0*DSP
    IF(G7.GT.XMFP)GO TO 720
710 P1=XMFP
    GO TO 730
720 IF(G7.GT.P1)GO TO 730
    P1=G7
730 S3=1.
    S2=1.
    F6=P1+B0
    F1=F6
    Z8=0.5
    IFL1=1
    GO TO 850
760 IF(P2.EQ.P3)GO TO 965


    IF(P2.GT.P3)GO TO 810
770 Z8=0.25
    IFL1=2
    GO TO 850
780 IF(P2.EQ.P3)GO TO 965
    IF(P2.GT.P3)GO TO 800
    Z8=-0.015
    GO TO 880
800 Z8=0.225
    GO TO 880
810 Z8=0.75
    IFL1=3
    GO TO 850
820 IF(P2.EQ.P3)GO TO 965
    IF(P2.GT.P3)GO TO 840
    Z8=0.475
    GO TO 880
840 Z8=0.72
880 IX1=0
881 IX1=IX1+1
    IIX1=IX1
    Z8=Z8+0.025
    IFL1=4
    GO TO 850
895 R(IX1,1)=P3
    R(IX1,2)=P2
    R(IX1,3)=Z8
    P5=P2-P3
    IF(P5.GE.0.)GO TO 930
    IF(IX1.LT.2)GO TO 2060
    GO TO 935
930 IF(IX1.LT.11)GO TO 881
935 S(1)=(R(IIX1-1,1)-R(IIX1,1))/(R(IIX1-1,3)-R(IIX1,3))
    S(2)=R(IIX1-1,1)-S(1)*R(IIX1-1,3)
    S(3)=(P(IIX1-1,2)-R(IIX1,2))/(R(IIX1-1,3)-R(IIX1,3))
    S(4)=R(IIX1-1,2)-S(3)*R(IIX1-1,3)
    S(5)=(S(2)-S(4))/(S(3)-S(1))
    Q1=S(5)*AOFP
965 F=Q1/AOFP
    IF(Q2.GE.Q1)GO TO 2135
    R4=RES-V1
    IF(R4.GT.0.)GO TO 990
    GO TO 1030
990 IF(G1.EQ.0.)GO TO 1010
    CALL  PRESS(A4,K4,RES,V1,Z4,U3,B5,T5,L4)
1000 H1=L4*B5
    GO TO 1015
1010 H1=K4*R4+A4
1015 B1=AOFP*((H1/RSIP)**(2*NO))
    V9=AOFP*(B1+AOFP)*0.5*U7*F*ORF
    IF(V9.GT.V1)GO TO 1210
1030 K6=K4*AOFP*0.5*U7*F*ORF
    I7=0
1040 I7=I7+0.1
    IFL2=1
    GO TO 1190
1050 IF(K8.GT.K6)GO TO 1040
    I7=I7-0.11
1060 I7=I7+0.01
    IFL2=2
    GO TO 1190
1070 IF(K8.GT.K6)GO TO 1060
    I5=I7
    I6=I7-0.01
```

```
       I7=I5
       IFL2=3
       GO TO 1190
 1095  K9=K8
       I7=I6
       IFL2=4
       GO TO 1190
 1110  K0=K8
       M=(K9-K0)/(I5-I6)
       B6=K9-M*I5
       I8=(K6-B6)/M


       H2=I8*RSIP
       V5=AOFP*0.5*F*ORF*U7*(1.+(H2/RSIP)**(2*NO))
       R4=RES-V5
       IF(G1.EQ.0.)GO TO 1165
       CALL PRESS(A4,K4,RES,V1,Z4,U3,B5,T5,L4)
 1155  H1=L4*B5
       GO TO 1170
 1165  H1=K4*R4+A4
 1170  B1=AOFP*((H1/RSIP)**(2*NO))
       V9=(B1+AOFP)*0.5*U7*F*ORF
       V1=V9
 1210  IF(RES.GT.V1)GO TO 1220
       V1=RES
 1220  K5=V1/(AOFP*U7)
       CALL COMFAC(F6,U1,B5,T5,Z)
       Z6=Z
       R3=(2.6994*G*F6)/(Z6*U1)
       VV=(5.62*((67-P3)**0.25))/(R3**.5)
       AA=0.7854*((D4/12.)**2-(D3/12.)**2)
       Q4=(3060*F6*VV*AA)/(Z6*U1)
 1780  P7=RSIP
 1180  WRITE(6,565) ILY,P7,AOFP,V1,K5,Q1,F6,Q4
 1805  V2=V2+V1
       IF(V1.EQ.RES)GO TO 2050
       IF(G1.EQ.0.)GO TO 2020
       CALL PRESS(A4,K4,RES,V1,Z4,U3,B5,T5,L4)
 1825  CONTINUE
 2010  RSIP=L4*B5
       GO TO 2025
 2020  RSIP=K4*(RES-V1)+A4
 2025  AOFP=AOFP*((RSIP/P7)**(2*NO))
       RES=RES-V1
       ILY=ILY+1
       GO TO 2040
  850  P2=RSIP*SQRT(1.-Z8**(1./NO))
       Q1=Z8*AOFP
       Q=Q1
       CALL BHFP(U0,U1,U2,U3,U4,F1,B5,T5,L,H,D,Q,S2,S3,K2,K3,IFLAG,F7)
       IF(IFLAG.GT.1)GO TO 710
  870  P3=F7
       GO TO (760,780,820,895),IFL1
 1190  IF(I7.GE.1.)GO TO 1205
       H2=I7*RSIP
       K8=(RSIP-H2)/(1.+(H2/RSIP)**(2*NO))
 1205  GO TO (1050,1070,1095,1110),IFL2
 2040  IF(IX4.LT.IYS)GO TO 611
       GO TO 2055
 2050  RES=0.
 2055  GO TO 2070
 2060  IF(P1.EQ.XMFP)GO TO 2070
       GO TO 710
 2070  CONTINUE
       WRITE(6,2150)V2
       WRITE(6,2155)RES
       GO TO 9999
 2120  WRITE(6,2125)
       GO TO 2070
 2135  WRITE(6,2140)Q2
       GO TO 2070
  480  FORMAT(3X,'BAROMETRIC PRESS : ',8X,F10.1,2X,'PSIA',/3X,'MINFLW PRE
      *SS      :',9X,F10.1,2X,'PSIG',/,3X,'ABANDONMENT BHP  :',9X,F10.1
      *,2X,'PSIA',/,3X,'RESERVES',9X,':',9X,F10.1,2X,'MMCF')
  500  FORMAT(3X,'FLW STRING LENGTH:',9X,F10.1,2X,'FT',/,3X,'FLW WELL HEA
      *D TEM:',9X,F10.1,'F',/,3X,'MID PERF DEPTH   :',9X,F10.1,2X,'FT',/,
      *3X,'STATIC W-HEAD TEM:',9X,F10.1,2X,'F',/,3X,'GAS GRAVITY      :',
      *9X,F10.1,2X,'AIR=1.',/,3X,'MOL FRACTION CO2 :',9X,F10.1,/,3X,'MOL
      *FRACTION  N2 :',9X,F10.1,/,3X,'MOL FRACTION H2S :',9X,F10.1,/,3X,'
      *CRITICAL PRESS   :',9X,F10.1,2X,'PSIA',/,3X,'CRITICAL TEMP    :',9
      *X,F10.1,2X,'R')
  535  FORMAT(3X,'PDF(P/Z/MMCF)      :',9X,F10.1)
  540  FORMAT(3X,'DSP FACTOR         :',9X,F10.1,/,3X,'RES TEMF         :',
      *9X,F10.1,2X,'F',/,3X,'AVG SLOPE(N)      :',9X,F10.1,/,3X,'INITIAL R
      *EDUCTION:',9X,F10.1)
  550  FORMAT(/,3X,'YEAR',3X,'RESERVOIR',7X,'ANNUAL',5X,'PERCENT',3X,'MAX
      * RATE',4X,'WELL HEAD',4X,'MIN RATE TO',/,8X,'PRES',6X,'POT',5X,'V
```

```
      *OLUME',7X,'OF',18X,'FLCW PRES',4X,'LIFT LIQUID',/,8X,'PSIA',5X, 'M
      *CFD',6X,'MMCF',6X,'POTENTL',5X,'MCFD',8X,'PSIA',10X,'MCFD')
  565 FORMAT(I4,2X,F7.1,F8.0,F10.0,F11.3,F12.0,F12.1,2X,F12.0)
 2150 FORMAT('OCUMULATIVE PRODUCTION   ',F7.0,' MMCF')
 2155 FORMAT('0 REMAINING RESERVES   ',F7.0,' MMCF')
 2125 FORMAT('0 PWF(MIN),GT,PC')
 2140 FORMAT('0 Q(MAX),LE, ',F7.1,' MCFD')
 9999 STOP
      END
      SUBROUTINE BHFP(U0,U1,U2,U3,U4,F1,B5,T5,L,H,D,Q,S2,S3,K2,K3,IFLAG,
      *F7)
      REAL L,K2,I1,M1,I2,N1,M2,M3,K3,I3,N2,M4
 9998 CONTINUE
      IFLAG=1
 1490 IF(S3.GT.0)GO TO 1515
      Y1=U3
      Y3=U0
      U5=U4
      GO TO 1530
 1515 Y1=U1
      Y3=U3
      U5=U2
 1530 CALL COMFAC(F1,Y1,B5,T5,Z)
      Z1=Z
      A1=F1/(Y1*Z1)
      DO=(L/H)*(D*Q*0.001)**2
      I1=A1/(((A1**2)*.001)+S2*DO)
      M1=K2/(2*I1)
      DO 1665 IXX=1,101
      F2=F1+S3*M1
      IF(F2.GT.0.)GO TO 1590
      IFLAG=99
      GO TO 1775
 1590 POO=F2/B5
      IF(POO.LT.16.)GO TC 1600
      IFLAG=99
      GO TO 1775
 1600 TOO=U5/T5
      CALL COMFAC(F2,U5,B5,T5,Z)
      Z2=Z
      A2=F2/(U5*Z2)
      I2=A2/(((A2**2)*.001)+S2*DO)
      N1=I1+I2
      M2=K2/N1
      S1=M2*N1
      F4=F1+S3*M2
      IF(F4.GT.0.)GOTO 1655
      IFLAG=99
      GO TO 1775
 1655 IF(F4.EO.F2)GO TO 1670
      M1=M2
 1665 CONTINUE
 1670 M3=(K3-S1)/N1
      DO 1750 IYY=1,101
      F3=F2+S3*M3
      CALL COMFAC(F3,Y3,B5,T5,Z)
      Z3=Z
      A3=F3/(Y3*Z3)
      I3=A3/(((A3**2)*.001)+S2*DO)
      N2=I2+I3
      M4=(K3-S1)/N2
      F5=F2+S3*M4
      IF(F5.GT.0.)GOTO 1740
      IFLAG=99
      GO TO 1775
 1740 IF(F5.EQ.F3)GO TO 1755
      M3=M4
 1750 CONTINUE
 1755 D1=(3*K3)/(I1+4*I2+I3)
      F7=F1+S3*D1
      IF(F7.GT.0.)GO TO 1775
      IFLAG=99
 1775 RETURN
      END
      SUBROUTINE PRESS(A4,K4,R2,V1,Z4,U3,B5,T5,L4)

      REAL K4,L4,L2,M5
      DIMENSION C(20,3)
 1830 IF(A4.EO.0.)GO TO 1845
      L2=K4*(R2-V1)+(A4/Z4)
      GOTO 1850
 1845 L2=K4*(R2-V1)
 1850 TOO=U3/T5
      POO=0.
      DO 1885 IW=1,20
      POO=POO+1
      C(IW,1)=POO
      CALL COMFC1(POO,TOO,Z)
      C(IW,2)=Z
```

```
      C(IW,3)=C(IW,1)*B5/C(IW,2)
      IF(L2.GT.C(IW,3))GO TO 1885
      GO TO 1890
1885  CONTINUE
1890  POO=POO-1
      DO 1920 IW=1,10
      PCO=POO+0.1
      C(IW,1)=POO
      CALL COMFC1(POO,TOO,Z)
      C(IW,2)=Z
      C(IW,3)=(C(IW,1)*B5)/C(IW,2)
      IF(L2.GT.C(IW,3))GO TO 1920
      GO TO 1925
1920  CONTINUE
1925  POO=POO-0.1
      DO 1955 IW=1,5
      IW3=IW
      POO=POO+0.02
      C(IW,1)=POO
      CALL COMFC1(POO,TOO,Z)
      C(IW,2)=Z
      C(IW,3)=(C(IW,1)*B5)/C(IW,2)
      IF(L2.GT.C(IW,3))GO TO 1955
      GO TO 1990
1955  CONTINUE
1990  M5=(C(IW3,3)-C(IW3-1,3))/(C(IW3,1)-C(IW3-1,1))
      B1=C(IW3,3)-M5*C(IW3,1)
      L4=(L2-B1)/M5
      RETURN
      END
      SUBROUTINE COMFAC(P,T,PC,TC,Z)
      PR=P
      X=T
      RRT=TC/X
      RP=PR/PC
      A=0.06125*RPT*EXP(-1.2*(1.-RRT)**2)
      B=RRT*(14.76-9.76*RRT+4.58*RRT**2)
      C=RRT*(90.7-242.2*RRT+42.4*RRT**2)
      D=2.18+2.82*RRT
      Y=0.01
      DO 2 J=1,30
      IF(Y.GT.1.)Y=0.6
      F=-A*RP+Y*(1.+(Y*(1.+Y*(1.-Y))))/(1.-Y)**3-B*Y**2+C*Y**D
      IF(ABS(F)-1.0E-6)4,4,3
3     DFDY=(1.+4.*Y*(1.+Y*(1.-Y))+Y**4)/(1.-Y)**4-2.*B*Y+D*C*Y**(D-1.)
2     Y=Y-F/DFDY
4     Z=A*RP/Y
      RETURN
      END
      SUBROUTINE COMFC1(RP,RT,Z)
      PR=T
      RRT=1./RT
      A=0.06125*RRT*EXP(-1.2*(1.-RRT)**2)
      B=RRT*(14.76-9.76*RRT+4.58*RRT**2)
      C=RRT*(90.7-242.2*RRT+42.4*RRT**2)
      D=2.18+2.82*RRT
      Y=0.01
      DO 2 J=1,30
      IF(Y.GT.1.)Y=0.6
      F=-A*RP+Y*(1.+(Y*(1.+Y*(1.-Y))))/(1.-Y)**3-B*Y**2+C*Y**D
      IF(ABS(F)-1.0E-6)4,4,3
3     DFDY=(1.+4.*Y*(1.+Y*(1.-Y))+Y**4)/(1.-Y)**4-2.*B*Y+D*C*Y**(D-1.)


2     Y=Y-F/DFDY
4     Z=A*RP/Y
      RETURN
      END
//GO.SYSIN DD *
      ANURADHA SINHA          GREENFIELD          ROTHMORE ESTATE
      HOLADAY           SALT LAKE COUNTY     UTAH
      12.0      800.      1000.     8000.      8000.
7050.          60.        80.       220.        .65
      0.0        0.0        0.0       0.0       0.0
      2.41       0.0      3450.     3000.
       .75        1.0       10.        .75      0.99
      71.        20.0       1.0       3.0
      71.       200.       800.0      1.0
      72.       175.       800.0      1.0
      73.       150.       800.0      1.0
      150.      800.        .75
/*
//
```

## OUTPUT LISTING

```
      ANURADHA  SINHA              GREENFIELD              ROTHMORE  ESTATE
      HOLADAY                 SALT LAKE  COUNTY            UTAH

           PROGRAM :  GAS DELIVERABILITY  FORECAST
```

| | | |
|---|---|---|
| BAROMETRIC PRESS : | 12.0 | PSIA |
| MINFLW PRESS : | 800.0 | PSIG |
| ABANDONMENT BHP : | 1000.0 | PSIA |
| RESERVES : | 8000.0 | MMCF |
| FLW STRING LENGTH: | 8000.0 | FT |
| FLW WFLL HFAD TFM: | 80.0F | |
| MID PERF DEPTH : | 7050.0 | FT |
| STATIC W-HEAD TEM: | 60.0 | F |
| GAS GRAVITY : | 0.6 | AIR=1. |
| MOL FRACTION CO2 : | 0.0 | |
| MOL FRACTION N2 : | 0.0 | |
| MOL FRACTION H2S : | 0.0 | |
| CRITICAL PRESS : | 669.9 | PSIA |
| CRITICAL TEMP : | 375.0 | R |
| PDF (P/Z/MMCF) : | 0.3 | |
| DSP FACTOR : | 1.0 | |
| RES TEMP : | 220.0 | F |
| AVG SLOPE (N) : | 0.8 | |
| INITIAL REDUCTION: | 0.8 | |

| YEAR | RESERVOIR | | ANNUAL | PERCENT | MAX RATE | WELL HEAD | MIN RATE TO |
|---|---|---|---|---|---|---|---|
| | PRES | POT | VOLUME | OF | | FLOW PRES | LIFT LIQUID |
| | PSIA | MCFD | MMCF | POTENTL | MCFD | PSIA | MCFD |
| 71 | 3450.0 | 3000. | 200. | 0.183 | 2803. | 812.0 | 1535. |
| 72 | 3378.6 | 2907. | 175. | 0.165 | 2710. | 812.0 | 1535. |
| 73 | 3317.0 | 2828. | 150. | 0.145 | 2630. | 812.0 | 1535. |
| 74 | 3264.8 | 2762. | 150. | 0.149 | 2563. | 812.0 | 1535. |
| 75 | 3213.2 | 2696. | 150. | 0.152 | 2498. | 812.0 | 1535. |
| 76 | 3162.1 | 2632. | 150. | 0.156 | 2433. | 812.0 | 1535. |
| 77 | 3111.5 | 2569. | 150. | 0.160 | 2370. | 812.0 | 1535. |
| 78 | 3061.3 | 2508. | 150. | 0.164 | 2308. | 812.0 | 1535. |
| 79 | 3011.7 | 2447. | 150. | 0.168 | 2246. | 812.0 | 1535. |
| 80 | 2962.5 | 2387. | 150. | 0.172 | 2185. | 812.0 | 1535. |
| 81 | 2913.7 | 2328. | 150. | 0.176 | 2125. | 812.0 | 1535. |
| 82 | 2865.4 | 2271. | 150. | 0.181 | 2066. | 812.0 | 1535. |
| 83 | 2826.8 | 2225. | 150. | 0.185 | 2020. | 812.0 | 1535. |
| 84 | 2769.8 | 2158. | 150. | 0.190 | 1951. | 812.0 | 1535. |
| 85 | 2722.6 | 2103. | 150. | 0.195 | 1896. | 812.0 | 1535. |
| 86 | 2675.7 | 2049. | 150. | 0.201 | 1841. | 812.0 | 1535. |
| 87 | 2629.2 | 1996. | 150. | 0.206 | 1786. | 812.0 | 1535. |
| 88 | 2582.9 | 1943. | 150. | 0.211 | 1732. | 812.0 | 1535. |
| 89 | 2537.0 | 1892. | 150. | 0.217 | 1679. | 812.0 | 1535. |
| 90 | 2491.9 | 1842. | 150. | 0.223 | 1627. | 812.0 | 1535. |

```
CUMULATIVE  PRODUCTION      3075.  MMCF

REMAINING  RESERVES         4925.  MMCF
```

### BIBLIOGRAPHY

Interstate Oil Compact Commission: Manual of Back-Pressure Testing of Gas Wells, Oklahoma City (1972).

# INDEX

Air-fuel ratios, 127

Annular flow, gas well, 183, 184

AOFP (absolute open flow potential)
  and gas well deliverability computation, 210
  stabilized, for gas well, defined, 191

Archie method for water saturation values, 19–20, 22

Bottomhole static/flowing pressure of gas well. *See* Gas well, bottomhole static/flowing pressure of

Bottomhole static pressure values, subprogram for, 210

Bubble point pressure, 55, 57

Carbon dioxide. *See* $CO_2$

$CO_2$ flood performance and oil recovery, program to estimate, 141–146
  description of, 142
  example problem for, 143–144
  method for, 141
  output from, 146
  purpose of, 141
  source listing of, 145–146

Conformance factor, 113

Density porosity log, 19, 20

DSP factor in gas well deliverability computation, 210

Electric log combination data, 22

Fassihi, Gobran, and Ramey algorithm, 127

Fluid properties, hydrocarbons in place, and reserves, programs for, 1–68

Formation volume factors and material balance calculations, 55

Fuel, air-, ratios, 127

Gas, reservoir engineering for, 153–219

Gas-cap gas injection, 81

Gas-cap gas withdrawal, 81

Gas compressibility factor, subprogram for, 156
  for gas in place, 164, 176
  for gas well deliverability computation, 210
  for hydrocarbon fluid PVT properties computation, 5
  in oil performance prediction program, 82
  for stabilized AOFP, 192
  for static/flowing bottomhole pressure, 184

Gas compressibility factor and viscosity, program to compute, 155–161
  description of, 156–157
  example problem for, 157–158
  method for, 155–156
  output from, 161
  purpose of, 155
  source listing of, 159–160

Gas formation volume factor, 5

Gas gravity, 5

Gas in place, initial, determined for water drive reservoir, program to calculate, 163–173
  description of, 164–166
  example problem for, 166–169
  method for, 163–164
  output from, 173
  purpose of, 163
  source listing for, 170–172

Gas in place and recovery factor for abnormally pressured reservoir, program to determine, 175–182
  description of, 176–177
  example problem for, 178–179
  method for, 175–176

*(continued)*
output from, 182
purpose of, 175
source listing for, 180–181
Gas-oil relative permeability data, 81, 82
Gas-oil relative permeability ratio, subprogram for, 115
Gas-oil relative permeability values, program for estimating, 13–18
description of, 14–15
example problem of, 15–17
method for, 13–14
output from, 18
purpose of, 13
source listing of, 17
Gas performance prediction by production decline analysis, program for, 71–80
description of, 72–73
example problem for, 73–74
method of, 71–72
output from, 79
purpose of, 71
source listing of, 75–79
Gas viscosity, subprogram for, 156
for estimating, 192
for hydrocarbon fluid PVT properties computation, 5
in oil performance prediction program, 82
Gas viscosity and compressibility factor, program to compute, 155–161
description of, 156–157
example problem for, 157–158
method for, 155–156
output from, 161
purpose of, 155
source listing of, 159–160
Gas well, bottomhole static/flowing pressure of, program for determining, 183–190
description of, 184–185
example problem for, 185–187
method for, 183–184
output from, 190
purpose of, 183
source listing for, 188–190
Gas well, conventional back-pressure test for, converting to equivalent isochronal test data, program for. *See* Gas well, point-after-point test data for
Gas well, point-after-point test data for, converted to equivalent isochronal test data, program for, 201–208
description of, 202–203
example problem of, 203–204
method for, 201–202
output from, 207
purpose of, 201
source listing for, 205–206

Gas well, stabilized absolute open flow potential (AOFP) of, program to calculate, 191–200
description of, 192–193
example problem for, 194–195
method for, 191–192
output from, 199
purpose of, 191
source listing for, 196–198
Gas well deliverability/production forecast, program for, 209–219
description of, 210–212
example problem for, 212–213
method for, 209–210
output from, 219
purpose of, 209
source listing for, 214–218
Gobran, Fassihi, and Ramey algorithm for oil-displaced/volume-burned method for in situ performance, 127

Hall-Yarborough procedure, 155
Havlena and Odeh procedure, 163
Hydrocarbon fluid, PVT properties of, program for computing, 3–11
description of, 5–6
example problem of, 6–7
method of, 3–5
output from, 10
purpose of, 3
source listing of, 8–10
Hydrocarbon index, 19
Hydrocarbons in place, fluid properties, and reserves, programs for, 1–68

In situ combustion performance, oil, and recoveries using empirical correlations, program to estimate, 135–140
description of, 136–137
example problem for, 137–138
method for, 135–136
output from, 140
purpose of, 135
source listing of, 139
In situ combustion performance, oil, using oil-displaced/volume-burned method, program to determine, 127–133
description of, 128–129
example problem of, 129–130
method for, 127–128
output from, 132
purpose of, 127
source listing of, 131
IOCC (Interstate Oil Compact Commission) manual, 183, 209
Isochronal test data, use of, for determining stabilized AOFP for gas well, 191, 192,

*(continued)*
193. *See also* Gas well, point-after-point test data for
Isopach data, 31

Material balance, gas, 209
Material balance calculations and formation volume factors, 55
Material balance computations, reliability of, 83
Material balance equation
for initial gas in place of abnormally pressured reservoir, 175
in oil performance prediction program, 97, 99
Tracy form of, 113
Material balance method
and bubble point pressure, 55, 57
for gas-in-place estimate, 163
for initial oil-in-place determination for reservoir with initial gas cap, 47–53
for initial oil-in-place estimate for a solution gas drive reservoir, 41–46
modified Schilthuis, 81
Movable oil index, 19

Net pay thickness, 19
Neutron porosity log, 19, 20

Oil formation volume factor, 5
Oil in place, initial, for partial water drive reservoirs with no gas cap, program for determining, 55–68
description of, 57–58
example problem of, 58–62
method for, 55–57
output from, 67
purpose of, 55
source listing of, 63–66
Oil in place, initial, for reservoir with initial gas cap, program for determining, 47–53
description of, 48–49
example problem for, 49–50
method for, 47–48
output from, 52–53
purpose of, 47
source listing of, 51–52
Oil in place, initial, for solution gas/depletion drive reservoir, program for estimating, 41–46
description of, 42–43
example problem for, 43–44
method for, 41–42
output from, 46
purpose of, 41
source listing of, 45–46

Oil in place and recoverable reserve, program for calculating, 31–40
description of, 33–35
example problem of, 35–37
method for, 31–33
output from, 39
purpose of, 31
source listing of, 38–39
Oil performance, prediction of, 69–124
Oil performance prediction and dispersed gas injection, program for, 113–124
description of, 115–116
example problem of, 116–118
method of, 114–115
output from, 123
purpose of, 113
source listing of, 119–122
Oil performance prediction by production decline analysis, program for, 71–80
description of, 72–73
example problem for, 73–74
method of, 71–72
output from, 79
purpose of, 71
source listing of, 75–79
Oil performance prediction and recovery of combination solution gas/gas-cap drive reservoir, program for calculating, 81–96
description of, 82–85
example problem for, 85–88
method for, 81–82
output from, 95
purpose of, 81
source listing of, 89–94
Oil performance prediction and recovery of a partial edge-water drive reservoir, program for, 97–111
description of, 99–100
example problem for, 100–106
method for, 97–99
output of, 110
purpose of, 97
source listing of, 107–110
Oil rates, estimates of, 127
Oil viscosity program, subprogram for, 5

Pay zone thickness, 19
Permeability. *See* Gas-oil relative permeability
Polymer flood performance and recovery, program to estimate, 147–152
description of, 148
example problem of, 149–150
method for, 147–148
output from, 152
purpose of, 147
source listing for, 151–152

Porosity, 19
Porosity logs, 19, 20
PVT data and oil performance prediction
    program, 81
PVT properties of hydrocarbon fluid. *See*
    Hydrocarbon fluid, PVT properties of

Ramey, Fassihi, Gobran, and, algorithm
    for oil-displaced/volume-burned
    method for in situ performance, 127
Ratio method of water saturation determi-
    nation, 20, 22
Recoverable reserve oil and oil in place,
    program for calculating, 31–40
    description of, 33–35
    example problem of, 35–37
    method for, 31–33
    output from, 39
    purpose of, 31
    source listing of, 38–39
Recovery, gas, program for
    example problem for, 178–179
    output from, 182
    source listing of, 180–181
Recovery, oil, $CO_2$ flood performance and,
    program to estimate, 141–143
    description of, 142
    example problem for, 143–144
    method for, 141
    output from, 146
    purpose of, 141
    source listing of, 145–146
Recovery, oil, and oil performance predic-
    tion of combination solution gas/gas-
    cap drive reservoir, program for, 81–96
    description of, 82–85
    example problem for, 85–88
    method for, 81–82
    purpose of, 81
    output from, 95
    source listing of, 89–94
Recovery, oil, and performance by empiri-
    cal methods, 202–52. *See also* In situ
    combustion performance, oil, and re-
    coveries using empirical correlations
Recovery, oil, polymer flood performance
    and, program to estimate, 147–152
    description of, 148
    example problem of, 149–150
    method for, 147–148
    output from, 152
    purpose of, 147
    source listing for, 151–152
Recovery estimates, oil, 127
Recovery factor, oil, calculating, 31
Regression analysis, 71
Reserves, fluid properties, hydrocarbons in
    place, and, programs for, 1–68

Reservoir
    closed volumetric, 41
    with fluid migration, 41
Reservoirs, programs for
    abnormally pressured, 175–182
    low-permeability, 191–200
    partial edge-water drive, 97–111
    partial water drive with no gas cap, 55–
    68
    solution gas/gas-cap drive, 81–96
    tight, 191–200
    water drive, 163–173
Reservoirs, solution gas drive, program for
    dispersed gas injection performance of,
    113–124
    gas-oil relative permeability values for,
    estimating, 13–18
    initial oil in place for, estimating, 41–46
Resistivity for program for conventional
    well log analysis, 19–20

Schilthuis material balance method, modi-
    fied, 81
Shaley sand, 19
Silts, 20
Solution gas-oil ratio, subprogram for, 5
Sonic porosity log, 19, 20
Stabilization of AOFP defined, 191
Standing-Katz $Z$-factor chart, 155

Three-porosity method, 19
Tracy form of material balance equation,
    113
Tubular flow, 183, 184
Two-porosity method, 19

U.S. DOE report, 135, 141, 147

Van Everdingen-Hurst
    unsteady-state equation of, 55, 56, 97
    values of, for water influx computation,
    163
Viscosity. *See* Gas viscosity
Volumetric method, use of, 31

Water influx
    and material balance equation, 47
    subprogram for, 165
Water saturation, 19, 22
Wellhead pressure, 210
Well log analysis, conventional, program
    for, 19–29
    description of, 22–24
    example problem of, 24–25
    method for, 20–22
    output from, 29
    purpose of, 19–20
    source listing of, 26–28